BLOCKCHAIN OR DIE *HARDER*

No Action Means No Profit

Eric L. Guthrie, Esq.

DEDICATION

Blockchain or Die Harder (here referred to as *Blockchain or Die*) is dedicated to Tracey Guthrie, MD, and Kate Guthrie, PhD, the best sisters in the world.

At the time of publication, Tracey, a psychiatrist by training, serves as Professor, Clinician Educator, Vice Chair for Clinical Faculty Affairs, Program Director of the General Adult Psychiatry Residency Program, and Assistant Dean for Diversity and Inclusion at the Warren Alpert Medical School at Brown University. Kate is a senior research scientist at the Center for Behavioral and Preventative Medicine at the Miriam Hospital and a Professor of Psychiatry and Human Behavior at the Warren Alpert Medical School of Brown University. She also holds a secondary appointment as a Professor of Behavioral and Social Sciences at the Brown University School of Public Health.

Both Tracey and Kate have achieved significant milestones in their careers, yet have never lost sight of the most important things in life: nurturing a happy marriage and going above and beyond to support the family. They have both been supremely supportive of all our parents, siblings, and their nephews, including Kevin Peter Guthrie, Jr., and Bryson Lance Guthrie, my son.

Their marriage stands as an admirable example of love, support, and communication. They have surrounded themselves with supportive friends and colleagues, people I would be proud to call "friends."

Initially, Tracey and Kate did not believe in cryptocurrencies, but they always supported all the members of our family and me in our professional and personal endeavors. As of 2025, they have become more open to the possibility of a world that uses cryptocurrencies. For that, and all the reasons mentioned above, I am proud to be their brother.

Thank you, sisters!

CONTENTS

FOREWORD

I met Eric in 2023 when speaking on Digital Assets at an event.

When Eric reached out about a year and a half later and asked me to read his new work, I knew that it was worth reading.

"Blockchain or Die Harder" is a valuable resource for anyone seeking to understand the world of blockchain and cryptocurrency. Each chapter provides a step-by-step approach that is both structured and engaging, avoiding the broad generalities found in many books on the subject. *Blockchain or Die Harder* is for those who want to engage with digital finance not only as consumers, but also as creators, innovators, and leaders.

This book speaks directly to businesses and individuals around the world who are ready to go deeper into the future of Digital Assets. Whether you already understand the basics, have been exploring blockchain, or are building something new in this space, this book will sharpen your thinking and expand your toolkit.

The Hon. David Burt, JP, MP

Premier of Bermuda

WHAT LEADING TYCOONS HAVE TO SAY ABOUT THIS BOOK!

“A Definitive Guide for the Blockchain Generation!

In today’s fast-moving digital landscape, clear and accessible resources on blockchain are invaluable. Blockchain or Die Harder by Eric Guthrie, Esq., is exactly that. As a college president, I view this work as an essential resource for anyone seeking to understand decentralized technology and its potential to shape the future.

What sets this book apart is its comprehensive scope. Guthrie skillfully navigates the world of cryptocurrencies and blockchain technology, offering clarity for both curious beginners and experienced professionals. His explanations balance depth with accessibility, covering everything from the foundational principles to real-world applications. The book is organized in a way that makes even complex topics easy to follow without sacrificing richness of content.

Where Guthrie truly shines is in his ability to translate technical language into everyday understanding. He uses relatable analogies that strip away the jargon, ensuring readers connect with the material. His “Scenarios” and “Lessons Learned” sections, in particular, are standout teaching tools that br*ing concepts to life with real-world relevance and practical lessons.*

For these reasons, I believe ***Blockchain or Die and Blockchain or Die Harder*** are not just excellent guides; they belong in classrooms. These works have the structure and depth to serve as college-level textbooks, equipping students with both theoretical knowledge and the practical insight needed to thrive in a decentralized future.

This book is not simply about technology—**it's about preparing the next generation of scholars and leaders to step confidently into a world where blockchain is no longer optional, but essential."**

— ***Chris Rey, JD***
President, Barber-Scotia College
Chairman, Council of Presidents, National Pan-Hellenic Council
Former International President, Phi Beta Sigma Fraternity, Inc.

"As CEO of EarthID, I have read many books on blockchain technology, but few have impressed me as much as Blockchain or Die Harder. It stands out for its blend of technical clarity, honest discussion of risks, global scope, and actionable advice. What truly sets it apart is its rare and insightful deep dive into digital asset scams and fraud, an often overlooked but critical issue. By addressing these risks head-on, Eric L. Guthrie provides an invaluable resource for building a more secure and trustworthy digital future. In a space crowded with hype and overpromises, Guthrie offers a grounded, structured path to get educated and take action, making this an essential read for everyone from technologists to policymakers."

— ***Priya Guliani***
CEO, EarthID

"'Blockchain or Die Harder' isn't just a sequel; it's a practical playbook for where crypto and blockchain are headed next. Eric Guthrie doesn't just explain the tech; he lays out the history, shows you the forces driving it, the scams to avoid, and the steps to take right now to use cryptocurrencies and blockchains. From policy shifts in D.C. to big cultural moments on the world stage, this book connects the dots in a way that's clear, real, and actionable. This balance of perspectives demonstrates that blockchain is far beyond just a tech niche, making it approachable for anyone still on the fence. The QR-coded action items and "Lessons Learned" sections are a great touch because they make it easy for readers to go from learning to actually using this new technology. Whether you are curious or serious about understanding this space and putting that knowledge to work, this is the guide you want in your hands."

— Marlon Williams

Founder of Starter Labs & Atlanta Blockchain Center

"Welcome to Blockchain or Die Harder, Eric Guthrie's insightful follow-up to the acclaimed Blockchain or Die. This new volume presents a comprehensive message about the importance of thoroughly researching where and how you invest before exploring cryptocurrency or other technology products and services. We always encourage our community to *'Do Your Own Research'*".

Decentralized communities face a fundamental question: where can reliable research tools be found, and how can the credibility of the information they produce be verified? Always verify with multiple resources. Hence, the creation of the Blockchain Legal Institute (a centralized library of decentralized resources at www.bli.tools) and the need to add this book to your library collection.

I have known Eric for many years, both as an attorney and a blockchain business entrepreneur. We both believe in the power of knowledge and empowering individuals to understand this new era of financial literacy. His books provide background knowledge to support the research needed to understand the trends we are seeing.

As a fellow author of The Bitcoin Cinderella blockchain adventure series, I understand the complexity of writing technical topics for a consumer audience. His book continues to serve as a must-read guide for businesses and individuals aiming to thrive in the evolving digital economy."

—Jacqueline Cooper, Esq.
MA LDT, NBCT, CCM, BMM
Co Founder of the Blockchain Legal Institute & Maryland Blockchain Association

INTRODUCTION

Life after *Blockchain or Die.*

Welcome to *Blockchain or Die Harder*. Writing *Blockchain or Die* ranks among the greatest accomplishments of my career. The book was a huge success, and I was honored when so many readers requested an updated or a second blockchain book.

Blockchain or Die achieved Amazon #1 New Release status, then #1 Bestseller status seven times. It also earned multiple honors:

- *2020 NYC Big Book Award Winner,*
- *2020 Independent Press Award Distinguished Favorite,*
- *2020 Indie Book Awards Finalist, and*
- *2020 American BookFest Finalist.*

These recognitions gave *Blockchain or Die* a great deal of exposure and established its position in the blockchain literature landscape.

Blockchain *or Die* was published by 'Better ME Better WE Publishing' in September 2019. Immediately after publication, I started a domestic book tour, which included book signings in Washington, D.C, Raleigh, North Carolina, Charlotte, North Carolina, and Atlanta, Georgia. Then, in November 2019, I began a global book tour with events in Bangkok, Thailand; Rio de Janeiro, Brazil; and Medellín, Colombia. My planned 2020 book signings included: Accra, Ghana; Johannesburg, South Africa; and Amsterdam, Netherlands. Then COVID-19 happened, and the rest of my in-person book signings, presentations, and conferences were canceled.

PROMOTING BLOCKCHAIN OR DIE DURING A GLOBAL PANDEMIC

After I published *Diversify or Die,* most of my revenue came from book sales at conferences and events. Conference attendees would line up to buy an autographed copy of *Diversify or Die*, providing a stable and steady growing stream of readers and revenue.

The global pandemic, however, virtually eliminated this in-person revenue stream. So, I had to adapt — of course, I was not alone in adapting my business to the realities of the global pandemic.

The COVID-19 global pandemic forced most industries to transition to a virtual work and business environment. Fortunately, *Blockchain or Die* deals with cryptocurrencies, which are sometimes called "virtual currencies." So, both my content and business model were well-suited for a transition from an in-person business environment to a virtual business environment. Moving from in-person to virtual presentations, seminars, and training allowed me to continue reaching audiences. I did my best to pivot, and the vast majority of *Blockchain or Die* book sales came in the form of e-books via Amazon.com.

This virtual environment also enabled me to train people from all around the globe from my home office. By embracing a virtual business model, *Blockchain or Die* received unprecedented international exposure, making it an even bigger success than my previous works. Parallel to this, my blockchain training also gained a bigger audience for a fraction of the travel cost. This experience reinforced my belief in the boundless potential of blockchain as an asset class to grow even in uncertain times, and highlighted a pressing need for training in this area.

AFTER THE PANDEMIC

The global pandemic has officially been over for two years now, and businesses continue to embrace the same virtual tools they used to stay afloat during the pandemic. Virtual meetings are now a common business practice, contactless, doorstep delivery of food and goods from online vendors is still prevalent, and companies continue to organize virtual conferences to attract sponsors and attendees. Today, companies combine the best of in-person and virtual technology and practices to enhance the future of technology and business.

ON BECOMING THE DIRECTOR OF TRAINING FOR THE GOVERNMENT BLOCKCHAIN ASSOCIATION

The Government Blockchain Association (GBA), a 501c(6) non-profit organization in Fairfax, VA, is where I began designing blockchain technology training programs. Serving as a leader in a global blockchain organization was helpful in building new connections and staying connected with fellow blockchain experts.

THE NEXT STEP: WRITING 'BLOCKCHAIN OR DIE HARDER'

During the pandemic, I started conceptualizing my third book. Based on the success of *Blockchain or Die* and the many requests for an updated book, I decided to write *Blockchain or Die Harder*. However, it is not a simple update to its predecessor; it is a comprehensive guidebook for readers with all levels of experience that delves deeper into the intricacies of blockchain technology, cryptocurrencies, and their future potential, incorporating new concepts and ideas.

HOW TO MAKE THE MOST OF 'BLOCKCHAIN OR DIE HARDER'

The thirteen chapters in '*Blockchain or Die Harder*' are designed for progressive learning and should be read in sequence. The important concepts presented in earlier chapters form the foundation for understanding the material presented later in the book.

The four parts and thirteen *Blockchain or Die Harder* chapters include:

Part I. The Current Landscape

Chapter 1: The Evolution of Cryptocurrencies

Chapter 2: The Crypto Evolution: Why Staying Updated is Key?

Chapter 3: Major Cryptocurrencies by Market Capitalization

Chapter 4: Regulatory Landscape for Cryptocurrencies

Chapter 5: Cryptocurrencies Around the World

Part II. Present Applications and Future Possibilities

Chapter 6: Blockchain Technology Updates

Chapter 7: The A–Z of Buying NFTs

Chapter 8: Is the Metaverse the Next Frontier for Everything?

Part III. Investing in Cryptocurrencies

Chapter 9: Should You Invest in Cryptocurrencies?

THE IMPROVED SUCCESS STRATEGY ACTION ITEMS

The Success Strategy Action Items section was one of the most highly praised parts of *Blockchain or Die*. When I completed my first draft of *Blockchain or Die*, I gave the draft copy to a few reviewers for feedback, including one of my former mentors. Like me, this mentor started investing in cryptocurrencies in 2017, so he had a fair amount of cryptocurrency experience before reading *Blockchain or Die*.

My mentor provided valuable feedback about the entire draft, but took special care to comment on the Success Strategy Action Items section. He read Chapter 1 to absorb the information and then implemented the success strategy action items himself. By the time he reached Chapter 3, he told me he was eagerly anticipating the Success Strategy Action Items recommendations. His feedback affirmed my belief that this section was a standout feature of *Blockchain or Die*.

Many *Blockchain or Die* Amazon reviews recognized the value of the Success Strategy Action Items. Here are a few excerpts of those reviews.

"Blockchain or Die contains a wealth of information on cryptocurrency and blockchain technology. Regardless of your level of knowledge or involvement, this book is a valuable resource and training tool. Best of all, each chapter

concludes with actionable "Success Strategy Action Steps." Get this book as a foundation for building your crypto and blockchain library."

"Blockchain or Die" is an excellent book, and the secret sauce is the "Success Strategy Action Items" at the end of each chapter. There are many books and articles on cryptocurrencies and blockchains, but Eric uses the action items to really engage readers. Eric uses the same educational interactive experience in Blockchain or Die as his blockchain training."

"I loved Eric's first book, Diversify or Die, and Eric's second book, Blockchain or Die, was even more informative. Diversity is an established business practice, but blockchain technology is new and can be intimidating. Blockchain or Die breaks down cryptocurrencies and blockchain technology, allowing me to grasp the fundamentals. Now, I am using the action items in the book to learn how to use this technology. As I learn more, I look forward to learning more.

"This is Eric's second book. Eric's first book, Diversify or Die, gave me a new perspective on diversity and inclusion. Eric's second book, Blockchain or Die, gave me a comprehensive introduction to cryptocurrencies and blockchain technology. For years, I had been hearing about blockchain technology, but I was so busy with work that I didn't have time to conduct my own research. I purchased Blockchain or Die. Eric explains both cryptocurrencies and blockchain technology extremely well and even gives the reader action items at the end of every chapter. Buy this book if you want to really learn how to use blockchain technology."

"I had a brief discussion with Eric Guthrie about blockchain technology. Eric took out Blockchain or Die and showed me a few important pages. I decided to purchase five copies of the book immediately. This information was too important to keep to myself, so I decided to buy four books for friends and family. After reading the book, I have a much better understanding of

blockchain technology and cryptocurrencies. I also started to try some of the action items. Blockchain or Die has opened up so many opportunities for me that I am going to read it again to ensure I didn't miss anything. Blockchain or Die is a must-buy book."

These are just a few of the reviews of *Blockchain or Die* on Amazon that reference the Success Strategy Action Items.

Of course, '*Blockchain or Die Harder*' has its own Success Strategy Action Items section. However, as I say in my speeches, "Organizations must evolve to stay relevant." I also consistently state, "Blockchain companies have to create products with a very accessible user interface to increase user adoption." And I must practice what I preach.

While *Blockchain or Die Harder* is not an actual blockchain product, its Success Strategy Action Items evolved to improve the user interface in a simple yet effective way. By 'evolution,' I mean that '*Blockchain or Die Harder*' contains QR codes. As you read each recommendation, you can use your smartphone to access the referenced resources immediately. This practical low-tech upgrade of QR codes provides easier access and faster action on the advice in this book.

SCENARIOS AND LESSONS LEARNED SECTIONS

The "Scenarios" and "Lessons Learned" sections are new additions to *Blockchain or Die Harder.*

Situational, real-life examples are valuable tools in adult learning. They provide a dynamic and engaging way to show readers the relevance of what they're reading, making the educational process more effective and enjoyable. Every scenario in this book is followed by at least one lesson learned, a concise takeaway with practical advice. Here are the benefits of the *Blockchain or Die Harder* scenarios and lessons learned:

1. Many scenarios are based on real-life cryptocurrency investment situations, allowing you to learn from others' mistakes.
2. Many of the scenarios encourage critical thinking before you make an investment. They are designed to help you develop important skills such as decision-making, critical analysis, and research techniques, all of which are essential in every aspect of cryptocurrency investment.
3. Cryptocurrency data and updates can be hard to apply personally. Scenarios help bridge this gap, turning theoretical knowledge into practical understanding for effective investment decisions.
4. The lessons learned sections highlight key takeaways. Mainly warnings on avoiding pitfalls and preventing mistakes that can cost you investment dollars.

THE TRUMP ADMINISTRATION AND CRYPTOCURRENCIES

The Trump Administration and Cryptocurrencies section is another new addition to *Blockchain or Die Harder.* The fact that Donald Trump is the 47th President of the United States is important because, during his campaign, he made powerful statements and campaign promises related to cryptocurrencies. During his Presidency, he enacted pro-cryptocurrency executive orders and appointed pro-cryptocurrency agency leaders. This book presents statements and policy decisions from President Trump that reflect his administration's cryptocurrency outlook.

SUMMARY

I hope this introduction has given you a clear idea of what to expect and piqued your curiosity about *Blockchain or Die Harder.* My goal is to deepen your understanding of cryptocurrencies and blockchain technology, and provide you with tools and resources to continue learning on your own. The more you engage with these strategies and tools in this book, the better equipped you will be to apply them to your cryptocurrency and blockchain journey.

No Action Means No Profit!

WHY BUY BLOCKCHAIN OR DIE HARDER? THE VALUE PROPOSITION

At every book signing, I observe several people pick up a copy of my book and flip through the first few pages. I am not always sure what they are looking for, but that cursory skim usually helps them decide— they either quickly put the book back down, start asking me questions, or buy the book. Instead of having potential readers flip through the pages to get a feel of the book, I created this Value Proposition section to summarize the value you can expect from reading *Blockchain or Die Harder.*

TEN CRITICAL LEARNINGS

The following chapters contain a wealth of information and practical guidance about blockchain technology and cryptocurrencies. Here's a snapshot of ten critical learnings in *Blockchain or Die Harder:*

1. Five ways to earn free cryptocurrencies.
2. The latest updates on cryptocurrency investment strategies.
3. Strategies to create a profitable blockchain business.
4. How to recognize and protect yourself from the latest scams.
5. How are corporations and countries adopting cryptocurrencies and blockchain technology.
6. What are NFTs, and how can you profit from them.
7. What the metaverse is, and how to profit from it.
8. Updates on cryptocurrency and blockchain legislation, regulations, and case law.
9. Insightful blockchain and cryptocurrency stories.
10. Of course, the ever-important Success Strategy Action Items!

These ten critical learnings will deepen your understanding of blockchain technology and cryptocurrencies.

COMMONLY ASKED CRYPTOCURRENCY QUESTIONS

Since publishing *Blockchain or Die*, I have been asked hundreds of questions about cryptocurrencies and blockchain technology. Once I started tracking the questions, I realized that certain doubts and questions cropped up repeatedly. *'Blockchain or Die Harder'* aims to provide a deeper understanding of blockchain technology and cryptocurrencies to a diverse range of readers — from laymen and beginners to seasoned investors and entrepreneurs. Therefore, in addition to the critical learnings above, Chapter 9 answers the following frequently asked questions:

1. Should I invest in cryptocurrencies?
2. Is it too late to invest in cryptocurrencies?
3. Which cryptocurrencies should I invest in?
4. Should I invest in Bitcoin or Altcoins?
5. How do I find the next "Dogecoin" before it takes off?
6. Should I invest in meme coins?
7. Should I stake my cryptocurrencies?
8. Is it too late to start mining cryptocurrencies?
9. What are the differences between Bitcoin and Ethereum in terms of investment potential?
10. What is DeFi (Decentralized Finance), and how can I participate in it?
11. How can I protect my cryptocurrency from hacks and scams?
12. What impact will central bank digital currencies (CBDCs) have on traditional currencies?
13. How can I use cryptocurrency for everyday transactions, and where is it accepted?

This value proposition is a preview of what you will learn in this book.

Are you excited?

I am! Let's get started.

Part I
THE CURRENT LANDSCAPE

CHAPTER 1
THE EVOLUTION OF CRYPTOCURRENCIES

"Last year [2023], we put out our bull case for Bitcoin. It was $1.5 million. With this institutional green light that the SEC [Securities and Exchange Commission] has provided, kicking and screaming though it did, the analysis we've done is that if institutional investors were to allocate a little more than 5% of their portfolios to Bitcoin, as we think they will over time, that alone would add $2.3 million to the projection I just gave you."[1]

Cathie Wood, CEO and CIO, Ark Invest

CHAPTER OVERVIEW

In 2009, the world saw the first functioning peer-to-peer digital money: Bitcoin. Built on the Bitcoin blockchain, it introduced a decentralized model of trust that eliminates the need for banks or governments. That breakthrough would later expand far beyond currency.

This chapter examines the evolution of cryptocurrencies, beginning with the introduction of Bitcoin, which sets the stage for the remainder of the book. It lays a foundation for a deeper understanding of cryptocurrency investment and future trends. For easier understanding, I divide the evolution of cryptocurrencies into four categories: Phase One, Phase Two, Phase Three, and Phase Four, and analyze key trends in each phase. If you are

---

[1] Pok, Caden. Cathie Wood Raises Her Bitcoin Prediction by $2.3 Million: What $1000 Invested Could Be Worth By 2030. Finance.Yahoo.com, April 30, 2024. https://finance.yahoo.com/news/cathie-wood-raises-her-bitcoin-141913532.html (Accessed October 28, 2025)

new to cryptocurrencies, this chapter serves as your key to understanding the background details that have contributed to the modern-day crypto space.

INTRODUCTION

As the CEO and CIO of Ark Invest, Cathie Wood is a prominent institutional investor with significant influence over the cryptocurrency market. Her quote is densely packed with insights about the past, present, and future of Bitcoin and cryptocurrencies.

Past – "Last year [2023], we put out our bull case for Bitcoin. It was $1.5 million."

Present – "With this institutional green light that the SEC [Securities and Exchange Commission] has provided, kicking and screaming though it did."

Future – "[T]he analysis we've done is that if institutional investors were to allocate a little more than 5% of their portfolios to Bitcoin, as we think they will over time, that alone would add $2.3 million to the projection I just gave you."

In addition, this quote references critical components of a potential cryptocurrency bull market, including the roles of institutional investors and the SEC regulations. In a few simple lines, Cathie Wood highlights several elements that will drive future cryptocurrency bull markets. However, one major element Woods did not mention: advancement and innovation. Advancement and innovation will be addressed in the next section.

ADVANCEMENT AND INNOVATION

Since 2009, cryptocurrency investors, miners, and users have witnessed a dramatic transformation in the cryptocurrency landscape. There have been significant technological advancements, new types of cryptocurrency products, creative innovations, regulatory developments, and, of course, market fluctuations. From the initial surge in popularity to the subsequent prolonged bear market, also known as "crypto winter," and the resurging bull market, cryptocurrencies have continued to captivate investors, technologists, media, and governments, to name a few.

Decentralized finance (DeFi) began with Bitcoin as the first decentralized digital currency. By operating without centralized control, Bitcoin was decentralized and functioned as a currency, laying the groundwork for other cryptocurrencies to develop more advanced financial instruments, including decentralized finance protocols. These protocols are applications built on blockchain technology that enable users to lend, borrow, and trade cryptocurrencies without intermediaries, revolutionizing the traditional financial system.

As a recap, Decryptionary defines "blockchain" as a digital ledger used to prove that a group of people reached an agreement about something. Blockchain records are permanent and very secure, preventing manipulation. The first example of blockchain technology is the initial records of Bitcoin transactions[2]. With DeFi, anyone with internet access worldwide can download a wallet to lend, borrow, send, or trade blockchain-based assets without relying on traditional financial institutions.

DeFi operates on a blockchain network, providing a secure and transparent platform for financial transactions. The backbone of DeFi is smart contracts — self-executing programs with terms directly written into code. These contracts automate processes, eliminating the need for intermediaries such as banks.

For example, users can lend their cryptocurrencies to earn interest or borrow funds by providing collateral. In a nutshell, many cryptocurrencies

---

[2] Blockchain. Decryptionary.com, October 9, 2017. https://decryptionary.com/glossary/blockchain/ (Accessed September 30, 2025)

offer the same services as traditional banks — lending, borrowing, and trading a currency. The primary difference is that banks are centralized, whereas many cryptocurrency services are decentralized.

DeFi platforms also offer decentralized exchanges (DEXs), where users can buy and trade various cryptocurrencies, and some support derivatives such as futures and options. This allows users to speculate on cryptocurrency price movements. Unlike centralized exchange platforms, all transactions and data on DeFi platforms are publicly visible on the blockchain, ensuring transparency and accountability.

Users interact with DeFi protocols through decentralized applications (dApps) that connect to their crypto wallets. For example, a lending protocol might allow users to lend or borrow cryptocurrencies. Smart contracts verify the terms of the loan, ensuring that interest is calculated accurately and that funds are released automatically upon repayment.

Many DeFi platforms utilize tokenized assets, which are digital representations of real-world assets, such as stocks or bonds. These tokens can be traded on decentralized exchanges, providing liquidity and enabling various financial products. DeFi platforms have gained immense traction, attracting billions of dollars in total value locked (TVL) while offering innovative financial products. However, as with all financial instruments, risks remain.

First, DeFi platforms are still largely unregulated, which means their future is uncertain. Second, unlike the Federal Deposit Insurance Corporation (FDIC), which protects depositors' money up to $250,000 per depositor per bank in the event of a bank failure, the US government does not offer any protections or insurance for DeFi platforms. So, no government guarantee. Finally, DeFi platforms lack the liquidity of large financial institutions. This means that in the event of a panic sell-off, the DeFi platform could struggle to fulfill orders.

The emergence of non-fungible tokens (NFTs) is another innovation that has significantly impacted the cryptocurrency ecosystem. NFTs represent unique digital assets, ranging from artwork and collectibles to virtual real estate. In the past decade, the NFT market has exploded in popularity, attracting mainstream attention and driving substantial value.

The metaverse is another recent innovation in the cryptocurrency ecosystem, a virtual world where people interact with friends, work, play games, build a business, and even shop. Cryptocurrencies can be used to purchase digital assets within the metaverse, such as land, avatars, or goods. Cryptocurrencies are a major component of the metaverse infrastructure, ushering in a new era of digital interaction and commerce.

REGULATORY SCRUTINY

As an attorney heavily involved in the cryptocurrency space, I have observed that regulatory scrutiny of cryptocurrencies has intensified in recent years. Governments, legislatures, and financial regulators worldwide have been grappling with how to regulate cryptocurrencies. This regulatory uncertainty presents a significant challenge to cryptocurrency adoption, as legislatures can alter the legal status of cryptocurrencies at any time.

Despite these challenges and uncertainties, cryptocurrencies offer significant benefits that keep them relevant and powerful across many circles. Cryptocurrencies, especially decentralized ones, empower people by offering financial options to underserved communities and global populations without access to traditional banking systems. Anyone with a cell phone and an internet connection can participate in cryptocurrency transactions.

Cryptocurrencies also provide financial access to companies that would otherwise be unable to utilize traditional banking and funding sources for their business operations. They are rapidly reshaping how financial transactions, data security, and organizational governance operate.

PHASE 1: THE EARLY YEARS (2009 - 2016)

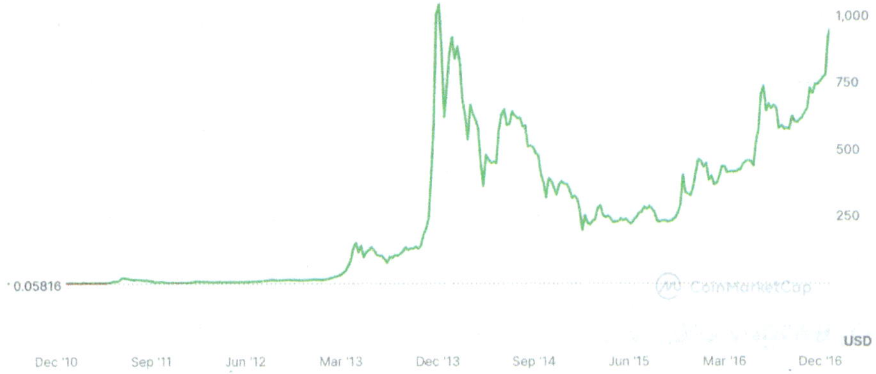

This chart offers a snapshot of Bitcoin's price evolution since its creation in 2009. As the chart shows, Bitcoin did not experience a significant price increase during its first three years. Although a price surge only began in 2013, the early years of Bitcoin and cryptocurrencies were truly revolutionary. I categorize these early years into four categories: Excitement, Challenges, and Limitations; Early Regulatory Challenges; Growing Community and Adoption; and the Development of Altcoins.

EXCITEMENT, CHALLENGES, AND LIMITATIONS

The early years of cryptocurrencies, following the introduction of Bitcoin, were marked with excitement and challenges. I learned about Bitcoin in 2016 and became fully involved with cryptocurrencies as an investor, user, attorney, consultant, and trainer in 2017, joining the scene at the tail end of the early years of the cryptocurrency movement.

Cryptocurrencies had not yet broken into the mainstream at that time. However, they already had a dedicated group of people who staunchly believed in the limitless potential of this new, decentralized asset class.

However, despite its early promise as an inclusive and accessible decentralized financial revolution, the cryptocurrency movement faced significant challenges and limitations, particularly in terms of volatility and scalability. Many people thought Bitcoin and cryptocurrencies were a fad or even a Ponzi scheme. During this initial phase, cryptocurrencies introduced business cases and technological advancement, but they had yet to produce any actual products, revenue, or profit. In the early years, people invested in cryptocurrency's potential rather than traditional metrics like income statements, balance sheets, or cash flow. Another major limitation was user adoption. As an attorney and consultant, I consistently made sure my clients considered this crucial point. Cryptocurrency companies can have the best idea in theory, but would individuals and businesses actually adopt or use their products or services? In the early years, using cryptocurrencies wasn't easy because the required technical knowledge was beyond the average person's grasp.

However, since then, cryptocurrency companies have developed easier user interfaces for the average person. These range from improved websites to mobile apps with back-end technology that enable users to simply point, click, or swipe to complete cryptocurrency transactions. These advancements, combined with the global expansion of internet access, have gradually brought cryptocurrencies from a purely technical discourse into the mainstream.

EARLY REGULATORY CHALLENGES

Drawing on my decades of legal experience, I focused on how governments, legislatures, and regulatory bodies struggled to regulate this new asset class. In fact, most of them are still navigating it. However, in the early years, there was considerable concern that government regulations would prevent the adoption or use of cryptocurrencies. Fortunately, most governments have adopted a balanced approach to dealing with cryptocurrencies, even as the regulatory landscape continues to evolve.

GROWING AND DEDICATED COMMUNITY

One benefit of being involved early with cryptocurrencies was joining a small but dedicated community committed to the success of these currencies. These communities became active worldwide, spreading the word about Bitcoin and cryptocurrencies. In my opinion, this was one of the main reasons why Bitcoin was successful — it represented a truly novel and liberated form of currency, free from establishment control. I do not think Bitcoin would have achieved the same success had it been started by a government or corporation. It was the dedicated grassroots efforts of programmers, financial experts, business professionals, and attorneys that gave Bitcoin and other cryptocurrencies a launchpad for success.

THE DEVELOPMENT OF ALTCOINS

The early years of Bitcoin, starting in 2009, were subsequently followed by altcoins—coins other than Bitcoin—beginning in 2011. The following table lists the first 20 cryptocurrencies chronologically:

Top Twenty Cryptocurrencies by Date of Introduction

NUMBER	YEAR	NAME	SYMBOL	2025 MARKET CAP
1	2009	Bitcoin	BTC	1
2	2011	Litecoin	LTC	20
3	2011	Namecoin	NMC	859
4	2012	Peercoin	PPC	1125
5	2013	Dogecoin	DOGE	9
6	2013	Gridcoin	GRC	Price Not Tracked

NUMBER	YEAR	NAME	SYMBOL	2025 MARKET CAP
7	2013	Primecoin	XPM	1736
8	2013	Ripple	XRP	3
9	2013	Nxt	NXT	Price Not Tracked
10	2014	Auroracoin	AUR	4436
11	2014	Dash	DASH	164
12	2014	NEO	NEO	128
13	2014	MazaCoin	MZC	Price Not Tracked
14	2014	Monero	XMR	22
15	2014	Titcoin	TIT	Price Not Tracked
16	2014	Verge	XVG	362
17	2014	Stellar	XLM	17
18	2014	Vertcoin	VTC	1317
19	2015	Ethereum	ETH	2
20	2015	Ethereum Classic	ETC	37

[3]

The above table highlights several noteworthy points. It clearly demonstrates that Bitcoin was the first cryptocurrency, and 16 years later, it is still the number one cryptocurrency by market capitalization. Furthermore, it highlights that although Ethereum was introduced six years after Bitcoin in 2015, it rose to the number two spot due to its business model and technological innovations, making it the second-largest cryptocurrency by market capitalization. Finally, although some cryptocurrencies were introduced very early (Namecoin – 2011, Primecoin – 2013, Nxt – 2013, Auroracoin – 2014, and Vertcoin – 2014), they were unable to compete with other cryptocurrencies introduced at the same time or later.

---

[3] List of Cryptocurrencies. En.wikipedia.org. https://en.wikipedia.org/wiki/List_of_cryptocurrencies (Accessed October 28, 2025)

The most important takeaway here would be the significance of newer developments and altcoins. If Bitcoin were the only cryptocurrency in existence, the world would lack new cryptocurrency business models to enhance healthcare, identity management, supply chain management, government services, energy efficiency, and cloud storage. Altcoin developers designed business models to address the issues and challenges in these industries.

PHASE 2: THE MAINSTREAM SURGE (2017 - 2019)

The above graph illustrates the impressive Bitcoin bull market from 2017 to 2019, during which Bitcoin reached a peak of $17,000 per coin. The 2017 cryptocurrency bull market was a period of unprecedented growth and excitement in the cryptocurrency space. This surge was fueled by a combination of factors, including increased mainstream awareness, speculative trading, and the emergence of new cryptocurrencies and blockchain-based projects. New and diverse investors, from individual retail traders to institutional funds, entered the market drawn by the potential for significant returns. After all, new money makes new market highs!

However, all bull markets eventually come to an end, and this one was no exception. In December 2017, Bitcoin and altcoins entered a bear market that lasted until December 2018.

I put the mainstream surge years into four categories: Advancement and Innovation; Initial Coin Offerings; Media Hype and Celebrity Endorsements; and Regulatory Developments and Challenges.

It is worth noting that these categories are not absolute. After practicing cryptocurrency and blockchain law, consulting, and training during this phase of mainstream adoption, I quickly realized how these categories were fluid, blending into a single, long causal loop chain. For example, a cryptocurrency with a new technology idea might decide to use the ICO process to raise capital for its company.

To increase marketing exposure, they hire a PR firm to launch a massive media campaign and secure a celebrity endorsement to boost publicity. However, the cryptocurrency does not perform as expected, and unfortunately, most of the investors lose some or all of their money. Some investors subsequently file complaints with the SEC, claiming the ICO was an unregistered security. Unfortunately, this was a very common scenario during the mainstream surge phase.

ADVANCEMENT AND INNOVATION

For me, the advancements and innovations in the cryptocurrency industry formed the most fascinating aspect of this phase. I loved learning about new products and developments and exploring how they could be applied to social and business environments. I would read white papers to see how companies framed the deployment of their products and how it would help social infrastructures or enterprises. It was like being on the cutting edge of technology and business.

The development of new blockchain technologies and applications, such as decentralized finance (DeFi) and non-fungible tokens (NFTs), expanded the potential use cases for cryptocurrencies and attracted a wider range of investors. In turn, new investors, with their new investor money, pushed the price of cryptocurrencies even higher. However, the impact of advancement and innovation extended far beyond the realm of actual technology. "Advancements" and "innovations" were powerful words in pitch decks to convince investors that they were investing in the "next Bitcoin" or "next Ethereum." During the mainstream phase, these became buzzwords to raise capital in initial coin offerings.

INITIAL COIN OFFERINGS (ICOS)

The rising popularity of ICOs has significantly shaped the mainstream cryptocurrency investment landscape. An ICO is a funding mechanism that enables projects to raise funds by issuing tokens. Many ICOs have raised millions of dollars, attracting investors eager to be part of the next big technological breakthrough and to earn huge profits.

The period between 2017 and 2019 was marked by significant growth and volatility in the cryptocurrency market, leading to increased regulatory scrutiny and challenges. The popularity of ICOs surged during this phase, with many projects raising substantial funds. However, the lack of regulatory oversight led to numerous scams and fraudulent offerings. Governments and financial regulators worldwide began scrutinizing cryptocurrencies and ICOs, raising widespread concerns about money laundering, tax evasion, and investor protection. By 2018, the SEC intensified their investigations into ICOs, and by 2019, the number of ICOs dramatically decreased.

As an attorney in the cryptocurrency and blockchain space, I was approached by many cryptocurrency start-ups pitching their ICOs to me with the intent of obtaining my investment. Nevertheless, I chose not to invest in any ICOs for two reasons: first, their revenue projections were never supported by verifiable data, but rather by assumptions of broad cryptocurrency adoption, which they used to extrapolate their project's success.

Secondly, none of the companies that pitched to me actually registered their ICO as a security with the SEC. Their lack of compliance raised a big red flag in my mind.

MEDIA HYPE AND CELEBRITY ENDORSEMENTS

Media hype and celebrity involvement were another critical part of Phase 2. The media played a crucial role in fueling the surge in both interest and price, with numerous articles and news segments highlighting the potential of cryptocurrencies to revolutionize finance and disrupt traditional institutions. Moreover, many companies purchased premium media coverage to market their cryptocurrency. I started noticing that well financed cryptocurrency companies heavily used social media to market their cryptocurrencies and ICOs. This was where celebrity endorsements became most prevalent. In many marketing campaigns, celebrities were paid to post comments, photos, or even videos promoting a specific cryptocurrency on social media. Examples of celebrity endorsement include:

Sports

- Floyd Mayweather
- Mike Tyson
- Cameron and Tyler Winklevoss

Music

- Mel B
- Nas

Television/Film

- Ashton Kutcher
- Jamie Foxx
- Gwyneth Paltrow

These celebrities have millions of followers on social media, guaranteeing massive exposure for an unknown cryptocurrency. Many loyal fans relied heavily on their endorsement, transforming cryptocurrencies from a relatively niche industry into a rapidly growing investment industry.

PHASE 3: THE COVID-19 PANDEMIC YEARS (2020 - 2024)

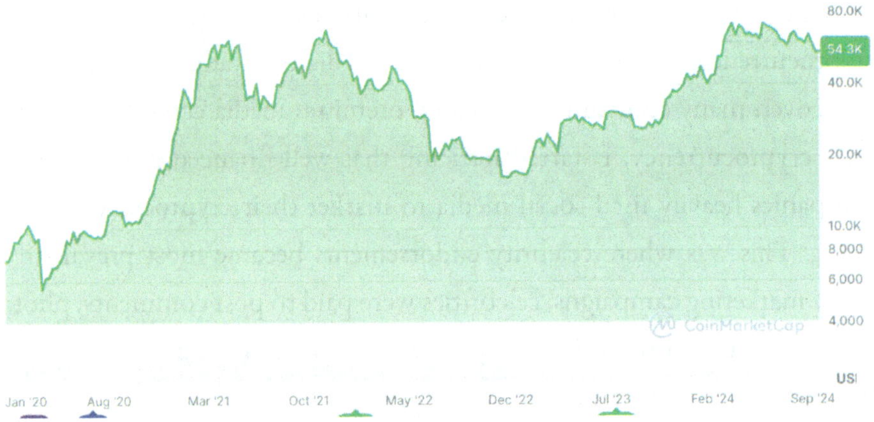

The above graph illustrates the dramatic rise of Bitcoin and the cryptocurrency bull market from 2020-2024, during which Bitcoin reached a high of $60,000 per coin. However, this phase also saw a major bear market as Bitcoin dropped from a high of $60,000 to $18,000. It took two years for Bitcoin to recover to the $60,000 price. This journey is best understood in two distinct parts: "The COVID years" (2020 – 2022) and the "COVID recovery years" (2022 – 2024). The remainder of this section will be divided into four categories: Market Maturity; Institutional Investors;

Advancements and Innovations; and Regulatory Developments and Challenges.

As part of Phase 3, it is worth noting that Bitcoin and several other cryptocurrencies have been around for over a decade, allowing them to mature as an asset class. Some of these cryptocurrencies had already established track records.

Year	Market Cap	% Change
2020	$757.31 billion	-
2021	$2.32 trillion	+67.35%
2022	$1.01 trillion	-56.4%
2023	$1.68 trillion	+118%
2024	$3.5 trillion	+108.33%

For example, staying on the CoinMarketCap.com Top Ten Cryptocurrencies by Market Capitalization for consecutive years brings credibility and attracts more serious, long-term investors. However, amid a semblance of growing stability and mainstream popularity, cryptocurrencies faced their first global challenge along with the rest of the global economy: *The COVID-19 pandemic*. Like many, I was genuinely curious to see how cryptocurrencies would respond to the global pandemic.

MARKET MATURITY

The COVID-19 pandemic significantly accelerated the maturation of the cryptocurrency market. Global upheaval and widespread economic uncertainty drove an increasing number of people, including myself, to seek alternative investments. Moreover, advancements in blockchain technology, such as enhanced scalability and interoperability, have improved user interfaces and increased the adoption of various cryptocurrencies. Finally, institutional investors and positive regulatory developments, though still evolving, have contributed to the maturation of the cryptocurrency market.

ADVANCEMENT AND INNOVATION

Apart from the direct impact of the pandemic, advancement and innovation in cryptocurrencies accelerated rapidly. New technological developments, such as Layer-2 scaling solutions and improved smart contract capabilities, enhanced transaction speed, reduced costs, and expanded the potential use cases for cryptocurrencies. In fact, I found that some Layer-2 solutions provided a much better user interface, which even everyday users (including myself) found Bitcoin payments both practical and appealing.

Bitcoin's arrival in 2009 proved that digital money could exist without a central bank. That was blockchain's first act: creating scarcity in a digital environment. However, Bitcoin had its limitations —users could mine and use Bitcoin, but not much else. Ethereum's launch in 2015 cracked the next barrier. For the first time, code could hold value and execute agreements: programmable money. A smart contract could automatically release escrow upon delivery confirmation — no lawyer, no middleman. These innovations sparked the rise of second-generation platforms like Cardano and Solana, which layered in staking economies and governance structures. Soon after, third-generation platforms like Polkadot, Cosmos, and Algorand targeted the hard problems: speed (sharding), interoperability (cross-chain bridges), and sustainability (proof-of-stake).

Blockchain technology was no longer a payments experiment; it was an expanding digital economy. The evolution of decentralized finance (DeFi) platforms was a major game-changer as it enabled users to access various financial services, including lending, borrowing, and trading, without relying on traditional intermediaries. Users have shared that they found the cryptocurrency lending process to be interesting and significantly faster and easier than traditional lending products. These combined developments positioned cryptocurrencies as a promising alternative to conventional financial systems.

INSTITUTIONAL INVESTORS

Institutional adoption was one of the defining stories of the COVID recovery phase. By February 2024, more than 600 firms had announced substantial investments in spot Bitcoin exchange-traded funds (ETFs) via their 13F filings submitted to the U.S. Securities and Exchange Commission (SEC). According to VirtualCapitalists.com, as of February 22, 2024, the following companies hold the most Bitcoins globally across public companies, amounting to a staggering $3.7 billion market capitalization:

RANK	COMPANY	COUNTRY	TOTAL BITCOIN	EST. VALUE AS OF FEB 22, 2024
1	MicroStrategy	US	174,530	$9.1B
2	Galaxy Digital	US	17,518	$912.1M
3	Marathon Digital	US	13,716	$714.1M
4	Tesla	US	10,500	$546.7M
5	Hut 8	Canada	9,366	$487.6M
6	Coinbase	US	9,181	$478.0M
7	Block Inc.	US	8,027	$417.9M
8	Riot Platforms	US	7,327	$381.5M
9	Hive Blockchain	Canada	2,596	$135.2M
10	CleanSpark	US	2,575	$134.1M
11	NEXON	Japan	1,717	$89.4M
12	Exodus	US	1,651	$86.0M

[4]

[4] Lu, Marcus. The Worlds Largest Holders of Bitcoin. TheVisualcapitalist.com, https://www.visualcapitalist.com/worlds-largest-corporate-holders-of-bitcoin/ (Accessed December 28, 2025)

This trend indicates increasing institutional confidence and integration of Bitcoin into mainstream investment portfolios, validating its legitimacy and accessibility as a viable asset class. The influx of institutional capital into these ETFs is a crucial step towards broader adoption and potentially greater price stability for Bitcoin. However, this is only the beginning, as discussed later in this book, other cryptocurrencies, including Ethereum and Cardano, are creating ETFs that will likely attract institutional investors.

REGULATORY DEVELOPMENTS

Of course, as an attorney, I paid close attention to global regulatory developments during Phase 3. As the cryptocurrency market matured and institutional investors committed more investment dollars, regulatory scrutiny also increased.

INCREASED REGULATORY SCRUTINY

- **Taxation:** Many countries implemented or clarified tax regulations for cryptocurrency transactions. This included reporting requirements, capital gains taxes, and VAT implications.
- **Anti-Money Laundering (AML) and Counter-Terrorism Financing (CTF):** Regulatory bodies worldwide strengthened AML/CTF rules for cryptocurrency exchanges and other businesses dealing with digital assets. This often involved requiring compliance with Know Your Customer (KYC) and Travel Rule standards.
- **Investor Protection:** Several jurisdictions have introduced measures to protect investors from scams and fraudulent activities involving cryptocurrencies. This included consumer education campaigns and licensing requirements for crypto platforms.

NOTABLE REGULATORY DEVELOPMENTS

- **United States:**
  - The Securities and Exchange Commission (SEC) increased its scrutiny of initial coin offerings (ICOs) and token sales, classifying many as securities subject to federal regulations.
  - The Financial Crimes Enforcement Network (FinCEN) issued guidance on the application of AML/CTF regulations to cryptocurrency businesses.

- **European Union:**
  - The European Commission published a proposal for a Markets in Crypto-Assets (MiCA) regulation, which aims to create a unified regulatory framework for cryptocurrencies within the EU.

- **United Kingdom:**
  - The Financial Conduct Authority (FCA) introduced a registration regime for cryptoasset firms operating in the UK.

EMERGING TRENDS

- **Central Bank Digital Currencies (CBDCs)**: Many central banks explored the development of CBDCs as an alternative to traditional fiat currencies through pilot projects and research initiatives.

It is important to note that the regulatory landscape for cryptocurrencies and blockchain continues to evolve rapidly. As the industry matures, we can expect further developments and potential challenges.

PHASE 4: THE TRUMP ADMINISTRATION AND CRYPTOCURRENCIES (2024 – 2029)

Blockchain or Die addresses legislation, regulations, and executive orders, not politics, but as elections go, the 2024 Presidential Election was a pivotal point for cryptocurrencies. As such, it is important to consider the impact of President Trump's administration on various aspects of cryptocurrency development.

President Trump has appointed several individuals to critical leadership positions in the government that have a substantial impact on cryptocurrency policy and regulation, including the Chairperson of the Securities and Exchange Commission, the Secretary of the Treasury, and the Chairperson of the Commodity Futures Trading Commission. All of these appointments, and their impact on cryptocurrencies, are discussed in detail in Chapter 4.

President Trump also created a new position in his administration: AI and Crypto Czar. If you are wondering what a "Czar" is and what it does for the President, as I was, here are the answers.

A Czar is a policy expert who advises the President and Senior White House officials in their area of expertise. Donald Trump selected David Sacks as the Artificial Intelligence (AI) and Crypto Czar to provide advice on artificial intelligence and cryptocurrencies. David Sacks is a venture capitalist, Former PayPal Chief Operating Officer, podcaster, and crypto investor and influencer.[5]

Finally, in March 2025, President Trump hosted the first Crypto Summit at the White House. This Crypto Summit is significant because it was the first time a President of the United States hosted a public cryptocurrency event at the White House.

---

[5] Canon, Gabrielle. Trump Picks Venture Capitalist David Saks as AI and Crypto 'Czar'. Theguardian.com, December 5, 2024. https://www.theguardian.com/us-news/2024/dec/05/trump-david-sacks-ai-crypto (Accessed October 28, 2025)

ADVANCEMENTS AND INNOVATION

As President Trump is still in the early stages of his presidency, it is too soon to report on actual technological advancements. However, cues to his technology mindset can be found in his close billionaire political allies from his 2024 Presidential campaign, including Elon Musk, President of Tesla, and Peter Thiel, Co-Founder of PayPal, Palantir Technologies, and Founders Fund.

The fact that Elon Musk and Peter Thiel, two technology billionaires, are part of the inner circle of Trump's administration could lead to greater focus on blockchain, cryptocurrency, and AI advancements.

CONCLUSION

This chapter briefly summarizes the journey of cryptocurrencies and draws a parallel to my own journey with the crypto industry — this is the first time I have reflected on the latter! My journey with cryptocurrencies has progressed through four meaningful phases, and I find that my experience reflects this pattern as well.

- **Phase One** – Discovery (2016 – 2018)
- **Phase Two** – Monetization (2019 – 2022)
- **Phase Three** – Scaling to Larger Projects (2022 – 2024)
- **Phase Four** – Determining the best approach to apply Phase Three to Phase Four based on a new regulatory framework (2025 – until)

Phase One was my research and learning phase. In hindsight, those years were akin to earning a master's degree in blockchain technology. I fully immersed myself in research, writing, and eventually, training. It was a time of gaining as much knowledge as I could.

In Phase Two, I needed to figure out how to monetize everything I had learned in Phase One; otherwise, all of my research and effort would merely remain a new hobby, not a profession. I had to start from scratch, learning how to acquire and serve new clients. Gradually, I created a client service model and used it as the foundation for my business. My initial clients were small start-ups, but eventually, I established a reputation as a blockchain technology consultant, trainer, and attorney.

In Phase Three, I began attracting larger clients with more secure funding. This is the phase I am in at the time of publication, focusing on delving deeper into client relationships in Phase Three and exploring the future potential of cryptocurrencies.

Phase Four is just getting started, and my approach is to analyze the new administration's regulatory stance toward cryptocurrencies and adjust my business model to maximize effectiveness and profit.

SUMMARY

This chapter sets the stage for the rest of the book by offering a concise background of the evolution of cryptocurrencies from the creation of Bitcoin in 2008 to the present market.

It also condenses information that will be explored in detail throughout the subsequent chapters. So, if something here piqued your curiosity or raised your investment antenna — great! You will find detailed follow-up information later in the book.

It is essential to note that the subsequent chapters analyze cryptocurrencies, CBDCs, and blockchain projects within the same chronological framework established in this chapter. While its trajectory remains uncertain, blockchain technology is already shaping finance, governance, and digital culture. The central question is not whether governments and institutions will adapt, but how quickly—and on whose terms. Chapter 2 explores how Bitcoin's origin story sparked this shift.

CHAPTER 2
THE CRYPTO EVOLUTION: WHY STAYING UPDATED IS THE KEY?

Cryptocurrencies are always evolving with new software and new ideas for implementation and adoption. Successful cryptocurrency investors research and understand the reasoning and impact of the evolutionary activity. Otherwise, they are investing based on yesterday's knowledge.

Eric Guthrie, Esq.
Author of *Blockchain or Die* (2019) and *Blockchain or Die Harder* (2026)

Blockchain or Die Harder is more than just a catchy sequel to Blockchain or Die — it is a warning that, as readers, investors, or even beginners, we must constantly stay updated on developments and advances in the cryptocurrency world if we want to utilize this asset to its fullest, limitless potential.

At this chapter's start, consider how cryptocurrencies have now become an integral part of global society. Bitcoin first traded at roughly $0.003. At the time of publication, it trades around $117,000, and the overall crypto market is worth over $3.7 trillion. This level of market capitalization does not occur without significant evolution, as numerous updates are required. Acquiring background knowledge to understand and invest in cryptocurrencies is no longer enough; staying current with crypto updates is critical.

CHAPTER OVERVIEW

In October 2008, as banks wobbled from the global financial crisis, a pseudonymous figure named Satoshi Nakamoto released a nine-page white paper to an email list. It described Bitcoin as a peer-to-peer currency system with no banks, governments, or central authority. On January 3, 2009, the first block was mined — with a message inside mocking bank bailouts."[6]

'Blockchain or Die Harder' covers the past, present, and future of cryptocurrencies and blockchain technology. This chapter begins that journey by analyzing:

- COVID-19's impact on cryptocurrencies
- The 2021 cryptocurrency crash,
- The Trump Administration and cryptocurrencies
- Important cryptocurrency updates.
- The presence of cryptocurrencies in pop culture and sports,
- Celebrities and cryptocurrencies,
- The Big Picture

ON ADOPTION & EARLY CRISES

Adoption of cryptocurrency followed strange paths. In 2011, the Silk Road marketplace accepted Bitcoin for a wide range of goods, from books to narcotics. The association was double-edged: it proved that Bitcoin worked as a form of money, but it also tied it to illegality. When the FBI shut down Silk Road in 2013, Bitcoin's price plunged — the first of many cycles where government action triggered extreme volatility. Yet outside the West, Bitcoin

---

[6] Sherry, Benjamin. What Is the Genesis Block in Bitcoin Terms? Investopia.com, August 28, 2024. https://www.investopedia.com/news/what-genesis-block-bitcoin-terms/#:~:text=our%20editorial%20policies-,What%20Is%20a%20Genesis%20Block?,address%20published%20on%20the%20blockchain. (Accessed October 29, 2025)

was also being used as a lifeline. In Argentina and Venezuela, citizens bought Bitcoin to escape crippling hyperinflation, while in Cyprus, bank freezes drove ordinary people to seek financial freedom through digital alternatives.

Another event happened in February 2014, Mt. Gox — at the time handling over 70% of Bitcoin trades worldwide — filed for bankruptcy after losing 850,000 BTC. For many outsiders, Bitcoin looked dead. But the collapse revealed an important truth: the failure was not in Bitcoin's code, but in the centralized exchange that managed people's coins. The lesson hardened the community's mantra: 'Not your keys, not your coins.'

By 2015, the first great governance battle emerged: should Bitcoin increase its block size to allow more transactions? The so-called 'Blocksize Wars' split the community, spawning factions and forks like Bitcoin Cash. The eventual compromise — Segregated Witness (SegWit) — was activated in 2017, proving that Bitcoin could evolve, but only after painful debates.[7] These disputes showed the world that Bitcoin was not just code, but a coordinated financial system.

COVID-19'S IMPACT ON CRYPTOCURRENCIES

The COVID-19 global pandemic had a profound impact on the global economy and financial markets, including the cryptocurrency sector. When the pandemic was in full effect, many cryptocurrency experts and I shared the same opinions on the ways the global pandemic affected cryptocurrencies:

1. The pandemic led to widespread economic uncertainty and market volatility. Many investors viewed cryptocurrencies as a hedge against inflation and risks associated with the traditional financial system.

[7] Segregated Witness. https://en.bitcoin.it/wiki/Segregated_Witness (Accessed November 2, 2025)

2. Governments around the world implemented massive stimulus packages to support their economies during the pandemic. This led to concerns about inflation and the devaluation of fiat currencies. Cryptocurrencies, not controlled by central banks, were viewed as potential safe havens.
3. The pandemic accelerated the shift towards remote work and digitalization. This shift made cryptocurrencies more accessible and appealing to a wider audience.
4. During lockdowns, staying at home saved lives but signified a dramatic shift in the everyday lives of most people. This stay-at-home requirement boosted online shopping, even for the basic necessities, which in turn sparked interest and investment in cryptocurrencies.

Overall, the pandemic created a favorable environment for the growth of cryptocurrencies, driving increased interest, adoption, and investment. However, it's important to note that the cryptocurrency market remains highly volatile, and investing in cryptocurrencies still carries significant risks, even when the market appears stable.

COVID'S IMPACT ON *BLOCKCHAIN OR DIE* SALES

The global pandemic also impacted my book sales. Before the COVID-19 pandemic, I sold many books in person at conferences and signings, using a credit card vendor account to accept payments. During the pandemic, I transitioned to selling *Blockchain or Die* e-books, accepting payments via digital apps and cryptocurrencies. However, when I realized I hadn't swiped a credit card to sell a book in over a year, I closed my vendor account.

As people attended meetings and training sessions virtually, cryptocurrency companies produced virtual content and training programs to educate them about the benefits, use cases, and investment opportunities of cryptocurrency. Again, the global pandemic forced me to adjust my marketing using webinars, virtual meetings, and podcasts. Before the pandemic, all my *Blockchain or Die* or my blockchain services training programs were in-person, not virtual. During the pandemic, over 90% of my services were by webinar, podcast, or virtual conference. All of these services were more cost-effective to conduct, as there were no travel costs for me, the presenter, or the attendees, and the training was completed in an environment conducive to the attendees. Virtual delivery cut costs, increased accessibility, and spiked attendance. It was not until 2024 that I transitioned back to in-person training.

This new virtual strategy led to a huge increase in sales of the *Blockchain or Die* e-book and helped it become an **Amazon #1 Bestseller** 7 times!

While the global pandemic presented numerous opportunities to utilize cryptocurrencies in everyday life, it also posed significant challenges for cryptocurrency companies, which were further compounded by other factors. One of them was the 2021 cryptocurrency crash.

THE 2021 CRYPTOCURRENCY CRASH

If you invested in cryptocurrencies before 2021, you experienced a major cryptocurrency event. In November 2021, most cryptocurrencies reached all-time highs before experiencing a spectacular crash.

On March 14, 2020, Bitcoin was trading around $5,200.00. Then, in November 2021, Bitcoin reached an all-time high of over $68,000. Many other cryptocurrencies also saw exponential price increases. However, on November 9, 2022, Bitcoin dropped to $15,883.16, and most other cryptocurrencies experienced a significant price decline.[8] So what caused the exponential price decrease?

Many factors contributed to the 2021 cryptocurrency crash, including the following:

De-pegging of TerraUSD (UST): UST is an algorithmic stablecoin developed by Terraform Labs that runs on the Terra blockchain. UST was pegged to the US dollar, but as an algorithmic stablecoin, it is not backed by any commodities or actual cash in the bank. TerraUSD marketed UST as an option to replace fiat-based global payment systems. Investors flocked to UST due to the 20% yield offered by Anchor, a cryptocurrency lender that supports UST holders. Large cryptocurrency investors recognized this as a short-selling opportunity. They borrowed substantial amounts of Bitcoin to buy UST, anticipating a significant price drop. In fact, short sellers contributed to the very decline they expected. This spooked investors, triggering massive sell-offs in UST and other cryptocurrencies.[9]

Inflation over 9% (percent): The pandemic caused inflation to increase to the highest rates in over a decade. Several factors contributed to the increase in inflation, including government stimulus, rising commodity prices, increased demand for certain goods, and labor wage pressures. All of these factors played a significant role in boosting inflation and disrupting the investment markets, including the cryptocurrency market.[10]

Federal Reserve interest rate hikes: Due to rising inflation, the Federal Reserve raised interest rates, which typically increases the prices of everyday utilities (such as food, gas, and clothes) and big-ticket items (such as houses

---

[8] CoinMarketCap.com. https://coinmarketcap.com/currencies/bitcoin/(Accessed September 28, 2025)

[9] Cole, Jordan. What Caused the Depeg of TerraUSD? An In-Depth Analysis of Its Collapse. BlockApps.com, December 2024. (Accessed October 28, 2025)

[10] Rugaber, Christopher. U.S. Inflation at 9.1 Percent, a Record High. PBS News, July 13, 2022. https://www.pbs.org/newshour/economy/u-s-inflation-at-9-1-percent-a-record-high (Accessed September 29, 2025)

and cars). When prices rise, individual investors have less money to invest and may sell investments to cover the higher cost of living. Investors, facing reduced disposable income, sold risky assets like cryptocurrencies, causing liquidity strains— a situation in which a company runs low or out of cash.

Reduction of venture capital spending: Venture capitalists exert significant influence over investment markets. When venture capitalists are investing in markets, individual investors take notice and follow. During the 2021 cryptocurrency crash, I personally observed many cryptocurrency projects that faced funding challenges as venture capital firms became more cautious about investing capital. The extreme price fluctuations experienced during the crypto winter made investors hesitant to fund cryptocurrency projects and businesses. Concerns about regulatory oversight and potential government restrictions also dampened investor sentiment. Finally, later in the cryptocurrency crash cycle, high-profile projects like Terra Luna and FTX further eroded venture capital confidence. When venture capitalists significantly reduce their investments in an industry, individual investors often follow suit, thereby increasing the risk of a market crash.

Of course, the cryptocurrency market rebounded after the 2021 crash, but it took years to recover. While the exact timing and pace of recovery varied for different cryptocurrencies, the market started an upward trend in 2023 and continued into 2024. By early 2025, Bitcoin had reached new all-time highs, and the overall cryptocurrency market capitalization had seen significant growth compared to the lows of the 2021 crash.

There are many more cryptocurrencies and blockchains than in 2019. For example, in 2019, only a handful of cryptocurrency exchanges existed; now, there are over 500 cryptocurrency exchanges worth billions of dollars. Additionally, global governments, Fortune 50 companies, venture capitalists, and angel investors are investing billions in research and development for cryptocurrency and blockchain technology. Simply put, individuals are making millions, and companies are making billions of dollars in the cryptocurrency market.

In 2024, cryptocurrencies are more than just code or investment statistics; they are an ever-growing part of society. In recent years, cryptocurrencies and blockchain technology have become integral to global business, popular culture, entertainment, and sports.

Now, considering the cryptocurrency market reached an all-time high in 2025, you might be tempted to think: "It's too late, I missed the boat!" The truth is, you absolutely haven't missed the boat! In fact, there is potential for many "boats" in the cryptocurrency world, and chances are the best "boat" is yet to be built. However, before investing in this technology, you must understand the evolution of cryptocurrencies, so you can decide your level of involvement.

THE TRUMP ADMINISTRATION AND CRYPTOCURRENCIES

During his campaign, President Trump met with cryptocurrency miners and executives, making cryptocurrencies a key part of his campaign platform.[11] During his first months in office, President Trump demonstrated his support for cryptocurrencies. President Trump's impact on cryptocurrencies will be covered throughout the book.

INDIVIDUAL CRYPTOCURRENCY UPDATES

This section will cover updates on four cryptocurrencies: Bitcoin, Ethereum, Ripple, and Cardano. Since 2020, cryptocurrency companies have implemented important updates to their mining and protocols. Often, when non-technical investors learn about upgrades, they may overlook them

---

[11] Seitz-Wald. Donald Trump Flips on Crypto: From 'Scam' to Building a 'Crypto Army'. NBC News.com, June 15, 2024. https://www.nbcnews.com/politics/2024-election/donald-trump-flips-crypto-scam-building-crypto-army-rcna157171 (Accessed October 26, 2025)

because they do not understand the upgrade. However, upgrades are critical to a highly functional and profitable cryptocurrency. Upgrades usually address the following technical issues:

- Fix bugs
- Improve speed
- Increase security
- Improve overall functionality, which may include new features

SUBSTANTIAL FINANCIAL IMPACT

Substantial financial impact is another type of cryptocurrency update. In the cryptocurrency world—such as Bitcoin halving events, hard forks, and the rise of new financial instruments like Exchange-Traded Funds (ETFs)—these events play a crucial role in shaping market dynamics. Substantial financial updates can directly or indirectly affect the price of the cryptocurrency.

The next section covers Bitcoin updates, including the Bitcoin halving, Bitcoin Exchange Traded Funds (ETFs), the eCash fork, and the Taproot upgrade.

BITCOIN HALVING

Bitcoin remains the number-one cryptocurrency by market capitalization and is the best-known. Although Bitcoin was conceived in 2008, its updates continue to drive the cryptocurrency market.

Bitcoin halvings are a crucial component of Bitcoin's economic infrastructure. As the chart below indicates, the halving occurs approximately every 4 years; when 210,000 blocks have been mined, the Bitcoin reward to miners is reduced by 50%. Initially, Bitcoin miners were rewarded with 50 Bitcoins for each successfully mined block; at the time of publication, they are rewarded with 3.125 Bitcoins.

Halving Date	Price (USD)	Block Height	Block Reward
1/3/2009	$.003	0	50 BTC
11/28/2012	$12.30	210,000	25 BTC
7/9/2016	$652.14	420,000	12.5 BTC
5/11/2020	$8,758.95	630,000	6.25 BTC
4/20/2024	$65,585.00	840,000	3.125 BTC

Bitcoin halving addresses the scarcity issue by reducing the supply of newly created Bitcoins over time. In other words, the halving minimizes the rate at which new coins are mined, ensuring that Bitcoin's overall supply remains limited to 21 million coins.[12]

In essence, Bitcoin halving is a key component in maintaining the value and stability of the cryptocurrency over the long term.

IMPACT ON BITCOIN SUPPLY

By halving the block reward, the rate at which new Bitcoins enter circulation is significantly reduced. This creates Bitcoin scarcity, which may drive up Bitcoin prices.

IMPACT ON BITCOIN PRICE

The reduced supply of new Bitcoins, coupled with a constant or even growing demand, can potentially lead to an increase in the price of Bitcoin. However, this is not guaranteed and depends on various market factors.

---

[12] Conway, Luke. Understanding Bitcoin Halving: Impact on Price and Investment Strategies. Investopia.com, August, 08, 2024. https://www.investopedia.com/bitcoin-halving-4843769 (Accessed September 28, 2025)

The anticipation of a halving event can lead to increased market volatility in the months before and after the halving. Investors might buy in anticipation of a price rise, and some might sell after the event if their price expectations aren't met.

Unlike fiat currencies, which can be printed at will, the supply of Bitcoin is fixed. This creates a deflationary environment where the value of each Bitcoin will increase as demand grows over time. The scarcity of Bitcoin also creates a sense of urgency and exclusivity. As demand increases, the limited supply can drive up prices. Bitcoin's scarcity makes it a potential store of value, similar to gold or other precious metals, providing investors with another option to hedge against inflation or market volatility.

THE INFLATION FACTOR

Inflation is always an important consideration in monetary policy. Although Bitcoin is not created or printed by any government, it had to address the issue of inflation to be effective as a currency. Otherwise, it would have the same problems as any other currency, and no one would use Bitcoin. Bitcoin halving addresses inflation control by controlling the Bitcoin supply to prevent excessive inflation, just like how traditional currencies can be devalued through excessive printing.[13]

The halving mechanism helps ensure that the finite supply of Bitcoins (capped at 21 million) is gradually released into circulation over time. This strategy contributes to the long-term decentralization of the network.

---

[13] Whittaker, Matt. Bitcoin Halving: How it Works and Why it Matters. Forbes.com, April 20, 2024. https://www.forbes.com/advisor/investing/cryptocurrency/bitcoin-halving/#:~:text=is%20at%20risk-%20Is%20Bitcoin%20Halving%3F,CEO%20of%20trading%20platform%20CoinRoutes (Accessed October 28, 2025)

IMPACT ON MINERS

After each "halving event," Bitcoin miners receive fewer Bitcoins for successfully validating transactions, which decreases their mining profits — the mining rewards are cut down by 50%, and Bitcoin prices are not likely to double, unless there is a drastic change in the market. Therefore, decreasing profits can cause miners to scale back or close their mining operations, especially those with less efficient mining hardware.

In conclusion, Bitcoin halvings are significant events that affect the network's economics and, potentially, the price of Bitcoin. Although it reduces miner profitability and may slow the issuance of new coins, it guarantees the long-term sustainability and security of the Bitcoin network.

BITCOIN SPOT ETF

In 2024, the Securities and Exchange Commission approved a Bitcoin spot ETF. "ETF" stands for "exchange-traded fund." A spot Bitcoin ETF is an exchange-traded fund that tracks Bitcoin's price by holding Bitcoin in reserve and backing each ETF share with a corresponding amount of Bitcoin. The term "spot" refers to the actual, immediate purchase and ownership of Bitcoin; therefore, the value of a spot Bitcoin ETF is based on the actual Bitcoin price. Unlike Bitcoin owners, purchasers of Bitcoin spot ETFs do not actually own Bitcoin — they are essentially purchasing a share of the ETF that represents their portion of the fund's Bitcoin holdings.[14] This helps investors gain exposure to Bitcoin and track its price without having to purchase and trade cryptocurrency in wallets or on exchanges.

The SEC's approval of Bitcoin spot ETFs is a significant sign of regulatory stability, suggesting a promising future for Bitcoin and other

---

[14] US Securities and Exchange Commission. "Statement on the Approval of Spot Bitcoin Exchange -Traded Products." January 10, 2024. https://www.sec.gov/news/statement/gensler-statement-spot-bitcoin-011023 (Accessed September 29, 2025)

cryptocurrencies. However, every investment carries risks. Before investing in a Bitcoin spot ETF, here are some pros and cons you must consider.

Bitcoin ETF Pros

- Investors can purchase the equivalent of Bitcoin through their regular investment accounts without having to set up a wallet.
- Investors can use the Bitcoin spot ETF in their existing retirement account.
- Investing in a Bitcoin spot ETF can offer tax advantages, as capital gains taxes may be deferred for long-term investors until the ETF is sold.

Bitcoin ETF Cons

- Bitcoin and other cryptocurrencies are a new asset class, which makes their future uncertain. The SEC can change or reverse its approval of the Bitcoin spot ETF.
- Investors must rely on the ETF issuer to accurately track and manage the actual Bitcoin holdings. If the ETF issuer encounters financial difficulties or fails to properly manage the ETF, investors could lose investment dollars.
- Like all ETFs, Bitcoin's ETFs charge management fees, which can reduce your returns.

Investors should consider the pros and cons before investing in an ETF.

THE ECASH FORK

A hard fork occurs when changes are made to the blockchain's protocol or programming layer, resulting in the creation of a new blockchain. It can be viewed as a set of upgrades to the blockchain software that create a new blockchain version incompatible with older versions, essentially splitting the blockchain or cryptocurrency permanently.

Since the Bitcoin Cash hard fork in 2017, Bitcoin has undergone several hard forks. The most recent Bitcoin hard fork was the eCash fork, which occurred on November 15, 2020. eCash (XEC) is a new version of Bitcoin Cash that is designed to be easier to use as electronic cash.

According to the e.cash website:

> eCash is a cryptocurrency designed for use as electronic cash. Just as the invention of email made it possible to send direct messages online, eCash enables the direct transfer of money to other people online. This includes being able to pay for goods and services.[15]

The eCash network enables upgrades without splitting the network, thereby preventing hard forks. This can help prevent the instability and confusion that often come with hard forks. While the benefits of the eCash system—faster transactions and lower fees—are separate from the lack of hard forks, the smooth upgrade process is a significant advantage for both users and developers.

However, eCash remains largely unknown as yet another form of Bitcoin. In fact, when I first learned of eCash, I thought it was an entirely new cryptocurrency. It was not until I researched eCash that I discovered it was a Bitcoin hard fork, and many of my colleagues reported the same initial reaction. Clearly, while it has the potential for widespread user adoption and disruption of traditional finance, there is considerable confusion and a lack of knowledge surrounding the eCash brand. It could take years, or even a decade, for the public or merchants to start using eCash.

---

[15] ECash FAQ. E.Cash https://e.cash/faq (Accessed September 29, 2025)

BITCOIN TAPROOT UPGRADE

Another significant Bitcoin update was the Taproot upgrade, which was implemented in November 2021. One of the main issues with Bitcoin was that transactions took an average of ten minutes to complete on the Bitcoin blockchain. Decentralized finance cryptocurrencies, after Bitcoin, including Litecoin, Stellar Lumens, and even Bitcoin Cash, have blockchains that can complete transactions in as little as a few seconds. To remain competitive, the Bitcoin community and miners upgraded the Bitcoin blockchain without creating a hard fork, as doing so would have created yet another version of Bitcoin.

The Taproot upgrade, which took seven years to complete, combines multiple signatures and transactions together. Taproot batching expedites the transaction process, allowing the Bitcoin blockchain to complete transactions more quickly.[16]

The Taproot upgrade received widespread support from the Bitcoin community and was activated through a consensus mechanism called Speedy Trial, which was designed to achieve a majority agreement among miners. Looking ahead, this precedent implies a pathway for future upgrades that could improve Bitcoin's privacy, scalability, and smart contract capabilities, thereby fostering continued adoption and potentially increasing its value. The Bitcoin community demonstrated an impressive ability to unite behind technical improvements, ensuring Bitcoin remains adaptable and robust in a rapidly evolving digital landscape.

I believe the Taproot upgrade was necessary for Bitcoin to remain relevant in the long term. Before the Taproot upgrade, the Bitcoin blockchain lacked the capacity to store NFTs, which are referred to as

---

16 Sharma, Rakesh. "Bitcoin's Taproot Upgrade: What You Should Know." Investopia, September 29, 2022. https://www.investopedia.com/bitcoin-taproot-upgrade-5210039#citation-1. (Accessed September 29, 2025)

ordinals on the Bitcoin blockchain. Now, the Bitcoin blockchain can store ordinals, which are similar to NFTs and are covered in detail in Chapter 7.

ETHEREUM SPOT ETF

As Ethereum is currently the number two cryptocurrency by market capitalization, I was not surprised by reports that the SEC was also asked to approve Ethereum ETFs. Similar to Bitcoin spot ETF owners, purchasers of Ethereum spot ETFs do not actually own ETH. Instead, they purchase a share of the ETF that represents their portion of ETH in the fund.[17]

As another spot ETF, the pros and cons of the Ethereum ETF are largely the same as those of the Bitcoin ETF mentioned above. However, we must note some additional pros and cons when considering the Ethereum ETF.

Ethereum ETF Pros

- Although Ethereum is the second-largest cryptocurrency by market capitalization, it lacks Bitcoin's global brand and exposure. The Ethereum ETF offers financial industry professionals the opportunity to market and sell the ETF to their clients.
- Ethereum is trading at a lower price than Bitcoin, which means investors can purchase more actual Ethereum ETF shares than Bitcoin ETF shares for the same or less amount of money.
- The launch of Ethereum ETFs could have a positive or neutral impact on the price of Ethereum. In the short term, there may be a boost in price due to increased demand from investors. However, the long-term impact remains unclear.
- Ethereum ETFs can potentially increase trading volume in Ethereum as they make it easier for investors to buy and sell the cryptocurrency.

---

[17] Torpey, Kyle. SEC Approves Spot Ether ETFs. Investopia, July 23, 2024. https://www.investopedia.com/sec-approves-spot-ether-etfs-8678873#:~:text=Key%20Takeaways,Optimism%20Around%20Spot%20Ether%20ETFs (Accessed October 28, 2025)

- The launch of Ethereum ETFs is arguably a step towards mainstream acceptance of Ethereum and the cryptocurrency market as a whole.

Ethereum ETFs Cons

- The price of Ethereum can be highly volatile, and this volatility will be reflected in the ETF's cost.
- The regulatory landscape for cryptocurrency is still evolving, and there is a risk that future regulations could adversely affect Ethereum ETFs.
- There is a risk that the custodian holding the underlying Ethereum for the ETF could default or be hacked.[18]

RIPPLE LAWSUIT UPDATE

As Ripple is currently the third-largest cryptocurrency by market capitalization, its actions have a significant impact on the overall cryptocurrency market. In December 2020, the Securities and Exchange Commission (SEC) filed suit against Ripple Labs, alleging the company did not register its cryptocurrency (XRP) with the SEC as a security.[19]

The SEC v. Ripple analysis is discussed in detail in Chapter 4. However, it is mentioned here because the lawsuit was one of the most significant developments in the cryptocurrency industry over the last five years. The summary judgment decision and the final decision clarified when a cryptocurrency should be considered a security, which would require the cryptocurrency to be registered with the SEC.[20]

---

[18] Pound, Jesse; Macheel, Tanaya. SEC Approves Rule Change to Allow Creation of Ether ETFs. CNBC, May 23, 2024. https://www.cnbc.com/2024/05/23/sec-approves-rule-change-to-allow-creation-of-ether-etfs.html (Accessed September 29, 2025)

[19] SEC v. Ripple Labs, Inc., No. 20-cv-10832, ECF No. 974 (S.D.N.Y. Aug. 7. 2024)

[20] Stempel, Jonathan. "US SEC Seeks to Appeal Ripple Labs Crypto Decision." Reuters, August 11, 2023. https://www.reuters.com/legal/us-sec-appeal-ripple-labs-court-ruling-2023-08-09/ (Accessed September 19, 2025)

CARDANO OUROBOROS

Cardano is a public, decentralized blockchain platform designed to provide a more secure, scalable, and sustainable infrastructure for decentralized applications (DApps) and smart contracts. Its native cryptocurrency is ADA.

The SEC is currently deciding whether to approve a Cardano ETF. In addition to the ETF, Cardano has been making significant updates in its development and ecosystem[21]. The latest updates include:

- **Chang Hard Fork**: This upcoming upgrade is a crucial step in Cardano's transition to the Voltaire era of decentralized governance. It will introduce features like liquid democracy and a constitutional republic, empowering the community to have a greater say in the network's development.
- **Hydra**: Cardano's layer-2 scaling solution, Hydra, is designed to exponentially increase the network's transaction throughput. It is expected to significantly improve Cardano's user experience and scalability.
- **Governance**: Cardano is working on establishing a large-scale decentralized autonomous organization (DAO) to manage the network's treasury and governance. This will further decentralize decision-making and empower the community.[22]

These updates are part of Cardano's ongoing commitment to innovation and growth.

Over 8,900 cryptocurrencies are currently listed on CoinMarketCap.com, which means there are a host of other crypto updates

---

[21] Huntley, Harvey. Cardano Price Prediction: SEC Postpones ETF Decision – ADA Could be Weeks Away From Huge Rally. CryptoNews.com, August 28, 2015. https://cryptonews.com/news/cardano-price-prediction-sec-postpones-etf-decision-to-november/ (Accessed September 29, 2025)

[22] Cardano. https://cardano.org (Accessed January 26, 2026)

that this chapter would be ill-suited to cover. The key message is to stay informed about all crypto developments, especially if you are a miner or an investor.

CRYPTOCURRENCIES IN SPORTS AND POP CULTURE

The stories behind cryptocurrencies — the good, the bad, and the ugly — have expanded beyond cryptocurrency companies, hackathons, and white papers. Over the last decade, several mainstream movies, TV shows, and sporting events have not only featured cryptocurrencies but also explored them as a central theme. Here are a few examples:

MR. ROBOT (2015 - 2019)

Mr. Robot is a television series that creatively addresses the complexities of hacking, corporate corruption, mental illness, and, of course, cryptocurrencies. The show follows Elliot Alderson, a brilliant cybersecurity engineer with social anxiety disorder and dissociative identity disorder, who aspires to change society.[23]

He is recruited by "fsociety," a clandestine hacktivist group led by the enigmatic Mr. Robot. Together, Elliot *and Mr. Robot* embark on a mission to bring down the powerful conglomerate E Corp.

With its intricate plot, compelling characters, incisive commentary on modern society, and incorporation of cybersecurity and cryptocurrencies, *Mr. Robot* was definitely ahead of its time. Notably, it is arguably the first TV series on a major cable network to introduce Bitcoin, cryptocurrencies, and CBDCs to viewers. Key episodes featuring cryptocurrencies, or central bank digital currencies (CBDCs), include:

---

[23] *Mr. Robot*, directed by Sam Esmail, USA Network.

- **S2 E3** - Ray Heyworth is speaking with Rat Tail (RT), the former Systems Administrator for Heyworth's dark web company, about the need to access the funds. RT says they keep emptying the Bitcoin wallets, and the server keeps crashing; he does not know how to stop it or migrate the system to set up a hot and cold wallet system. This is the first mention of Bitcoin and cryptocurrencies in the series.
- **S2 E4** - The eCoin strategy is introduced during a conversation between Whiterose, leader of the Dark Army, a Chinese hacker group, and Phillip Price, CEO of E Corp.
- **S2 E7** - The local TV news reports that eCoins are accepted by taxi drivers with instructions to download the eCoin app. Once you reach your destination, scan the QR code, and the payment will be withdrawn from your eCoin wallet. Ray uses Elliot to reconnect his Bitcoin wallet to his darknet site.
- **S2 E11** – Phillip Price lobbies Jack, a high-level Administration Official, to allow E-Corp to make eCoin loans. Jack responds, "That is unconstitutional; you can't make your own currency; that is the federal government's job." Price responds, "The problem here is that hard cash is fading rapidly. Bitcoin is spreading, and if Bitcoin takes over, we will all be in a world of hell. With eCoin, we control the ledger and the mining servers." Later in the same episode, the local TV news reported that eCoin was experiencing "soaring popularity," so the personal lobbying was successful.
- **S3 E2** – Phillip Price announces that eCoin is the global currency against which all others will be measured, except for China, which adopted Bitcoin.

At the time Season 2 aired, which introduced cryptocurrencies into the storyline, most people were not familiar with Bitcoin. In fact, in 2016, Bitcoin traded at a very low price range, between $433 and $963. The show not only told the mainstream audience about Bitcoin and cryptocurrencies but also highlighted the potential of digital assets to challenge established power structures.

More significantly, *Mr. Robot* presented cryptocurrencies as a viable alternative to fiat currencies, making them less intimidating for some viewers. The show's portrayal of cryptocurrencies as a tool for social and political change may have influenced hundreds of viewers to explore investing in or using digital assets. While that is a matter of personal speculation, it is undeniable that Mr. Robot provided viewers with a basic understanding of cryptocurrency concepts, such as blockchain technology, mining, and wallets, demystifying blockchain and generating interest among a wider demographic.

CRYPTO (THE MOVIE)

Crypto (2019) is a crime drama movie centered on cryptocurrency and money laundering. It creatively explores how cryptocurrency forensic investigations can help track cryptocurrency criminal activity. Martin, a young and talented auditor, along with Earl, a long-time friend, IT wiz, and cryptocurrency miner, uncovers a criminal syndicate with dealings in drug trades, terrorism, and human trafficking in a small town. Once the crypto dominoes begin to fall, havoc and mayhem follow.[24]

Crypto shows that cryptocurrency mining and crimes happen in all parts of the world, even in small rural towns where one might think crypto is not accessible. This is contrary to the misconception prevalent in the general perception of cryptocurrencies in 2019, when the movie was released. It also showcases the importance of cryptocurrency forensic investigations for maintaining compliance and integrity in the industry. The allure of profits

[24] *Crypto*, directed by John Stalberg Jr. (Lionsgate, 2019).

and the hype of celebrity endorsements are great for the cryptocurrency industry.

SUPER BOWL LVI (2022)

On February 13, 2022, during Super Bowl LVI between the Los Angeles Rams and the Cincinnati Bengals, cryptocurrency advertisers dominated the commercial breaks, turning the event into what many dubbed as the "Crypto Bowl".

I was watching with family and friends, prepared to enjoy Super Bowl LVI, a good ol' American pastime. Within fifteen minutes of kickoff, the first cryptocurrency commercial appeared. The Super Bowl was an exciting event because it would undoubtedly increase public interest and investment in cryptocurrencies through enormous exposure. However, what followed next was something I was unprepared for.

I saw a flood of cryptocurrency commercials for various companies. I was pleasantly surprised to see A-List celebrities, including LeBron James and Larry David, star in high-quality cryptocurrency commercials.

The Coinbase commercial was the most unique as it did not feature any actors, music, or a storyline. It simply displayed the Coinbase QR code floating around the screen. I immediately pulled out my smartphone to scan the QR code; however, the website was down due to a massive spike in traffic. This was the most powerful example of cryptocurrency truly becoming mainstream!

BITCON (THE MOVIE)

Bitcoin (2022) is an action-packed movie about Ty Adams, a Bitcoin miner and reformed criminal, and Jess, a clever young woman, searching for a missing hard drive holding $800,000 in Bitcoin. Dangerous criminals are also looking for the same hard drive. To improve their chances of finding the hard drive and staying alive, Ty and Jess make some dangerous alliances. [25]

In addition to a clever storyline, I liked two things about *Bitcon*: First, the movie displays definitions of cryptocurrency terms on screen, which then come into play in the storyline. These definitions are particularly helpful for viewers who may not be familiar with cryptocurrencies.

Second, as the movie progressed, the price of Bitcoin continued to increase, which was also displayed on screen. The increased price of Bitcoin added to the drama, as it raised the stakes for all parties seeking the hard drive. Although the film was a thriller that did not necessarily delve into the complexities of crypto, I believe its thriller tone would help demystify the world of cryptocurrencies for general audiences.

THE FIFA WORLD CUP (2022)

During the 2022 FIFA World Cup quarter-final match between Morocco and Portugal, I noticed the Crypto.com display on the field within the first minute. I kept seeing the Crypto.com display throughout the entire 2022 World Cup. In fact, Crypto.com was an official sponsor for the 2022 World Cup.[26] The World Cup is one of the premier global athletic events, and Crypto.com was front and center on the world stage. In the 2022 World Cup final, where Argentina played against France, the Crypto.com display remained front and center on the field.

While I didn't notice any cryptocurrency ads during the World Cup matches, Crypto.com still made a strong impression. Their display highlighted how cryptocurrencies have become a vibrant part of football

---

[25] Osterman, Matt. 2022. Bitcon. Propellent Films

[26] Company News. Crypto.com. March 22, 2022. https://crypto.com/company-news/crypto-com-unveiled-as-fifa-world-cup-qatar-2022-official-sponsor (Accessed October 28, 2025)

culture, even guiding millions on where to start: "Crypto.com." Another noteworthy aspect was catching the attention of a global audience. Although the Super Bowl is watched worldwide, its audience remains primarily American. The FIFA World Cup is truly an international sporting event; the fact that 32 nations qualified for the 2022 FIFA World Cup demonstrates the immense global interest. According to FIFA's report, over 1.5 billion people engaged with the 2022 World Cup.[27] Now that is exposure!

SUPER BOWL LVII (2023)

In Super Bowl LVII, the Philadelphia Eagles played against the Kansas City Chiefs. This time, I was looking for advertisements, but unlike Super Bowl LVI, Super Bowl LVII did not have any cryptocurrency ads. Since the cryptocurrency market suffered a significant decline in November 2021, resulting in billions of dollars in losses, many cryptocurrency companies opted against spending millions on Super Bowl ads during that time.

To me, the absence of cryptocurrency commercials in Super Bowl LVII carried an important message: when cryptocurrency is hot, everyone wants to be part of the action, but when the market is down, investors and advertisers stay clear, afraid to invest or spend money in the cryptocurrency market.

CRYPTO BOY (THE MOVIE)

In *Crypto Boy* (2023), Amir El Neni, a twenty-year-old immigrant whose Egyptian family owns a struggling Mexican restaurant in Amsterdam, Netherlands, attends a Cryptocore Capital sales and recruiting event. During the event, Roy, the Cryptocore CEO, tells the audience, "The only thing you have to do is join [Cryptocore]. Our traders do the rest. And we get you

---

[27] Inside FIFA. One Month On: Five Billion Engaged With FIFA World Cup Qatar 2022. InsideFIFA.com, January 18, 2023. https://inside.fifa.com/tournaments/mens/worldcup/qatar2022/news/one-month-on-5-billion-engaged-with-the-fifa-world-cup-qatar-2022-tm (Accessed October 26, 2025)

returns [of] one, two, sometimes three percent, I mean it. Per f**king day. Yes!"

Amir is so impressed by Roy's speech that he becomes a Cryptocore salesperson the next day. Roy pulls Amir into his lavish lifestyle, but neither Roy nor Cryptocore is what they seem. Roy's business practices and promises of wealth put Amir in a difficult position, forcing him to choose between Cryptocore and his family and friends.[28]

Crypto Boy reveals the dark side of cryptocurrency from several unique perspectives. First, it showed the anatomy of a cryptocurrency scam as the storyline unfolded. Roy is an intelligent, charismatic figure who exudes success, a description that also applies to the central figure in most scams.

After all, it takes a confident man (con-man) with intelligence, charisma, and the illusion of success to attract investors. Second, *Crypto Boy* shows the impact of a well-coordinated scam on a tightly knit immigrant community. Many immigrants invest their retirement savings in Cryptocore cryptocurrency, and when it crashes, their financial futures are jeopardized. This component is particularly pertinent in the real world — yet it is a perspective that is not often showcased in mainstream media, and the general public remains largely ignorant of the insidious and damaging nature of cryptocurrency scams.

SQUID GAME: SEASON 2 (2024)

Squid Game is a South Korean action, drama, and thriller series where an unnamed conglomerate aggressively recruits candidates in severe financial distress to participate in a series of children's games in small and large groups. The players are given a number to identify themselves, so the conglomerate can track them during the games. The winner (s) of the games receive 45.6

[28] El-Hamus, Shady. 2023. Crypto Boy. Caviar Films.

billion WRG (30.9 million USD). Without revealing too much, it is an understatement to say the games are very intense. [29]

Season 2 introduced Myung-gi Lee, also known as Player 333, a cryptocurrency influencer. Lee, formerly a prominent YouTuber behind the MG Coin crypto channel, finds himself in the game after going all-in on the Dalmatian coin, which crashes after its creators disappeared with all invested funds in a classic rug pull. To make matters worse, Myung-gi Lee also instructed all his followers to invest "every last penny," and violated several laws, putting a target on his back from both his followers and the authorities. He's depicted as bankrupt and a wanted man.

Desperate for money, Myung-gi Lee agrees to compete in the Squid Games, where he encounters some of the other players who followed his advice and lost money in Dalmatian coin. This is another story that has real-world applications. Be careful when listening to social media influencers' advice on cryptocurrencies. Conduct thorough research and consult with experts before making an investment. On that note, the next section delves into the intersection of celebrities and cryptocurrencies.

CELEBRITIES AND CRYPTOCURRENCIES

Celebrity involvement is another significant update that underscores how deeply cryptocurrency has become integrated into society, particularly in pop culture. Since 2017, numerous A-list celebrities have not only purchased cryptocurrencies but also invested in cryptocurrency companies and actively promoted specific cryptocurrencies through speeches, songs, advertisements, and social media posts. All of the following celebrities were early adopters and investors in cryptocurrencies because they saw huge potential for future growth and profit.

---

[29] Hwang Dong-hyuk, *Squid Game*, created by Hwang Dong-hyuk, Netflix, 2021, streaming video.

Sports

- **Floyd Mayweather:** Retired world champion boxer in multiple classes and self-promoter, rebranded himself as "Floyd Crypto Mayweather." He promoted the initial coin offerings (ICOs) of StockX.com and Hubii Network.[30]
- **Mike Tyson**: A former Heavyweight Boxing Champion and television and movie star, Tyson is another boxer who has entered the "cryptocurrency ring." In 2015, Mike partnered with Bitcoin Direct to introduce a Bitcoin ATM and a digital wallet for iOS and Android.[31]
- **Cameron and Tyler Winklevoss:** The Winklevoss twins, Olympic Rowers who competed in the 2008 Beijing Olympic Games, and also won Bronze at the 2009 World Cup in Lucerne, Switzerland. They were early Bitcoin investors, investing $11 million in Bitcoin in 2013, which reportedly made them Bitcoin billionaires.[32]

Music

- **Mel B:** A former Spice Girl singer, Mel B was the first musician to accept Bitcoin for her 2013 single "For Once in My Life."[33] To put her accomplishment in perspective, on January 4, 2013, Bitcoin was worth $13.50 USD.

---

[30] Chapparo, Frank. Floyd Mayweather is Promoting Another Initial Coin Offering: "You Can Call me Floyd Crypto Mayweather. August 23, 2017. https://www.businessinsider.com/floyd-mayweather-promoting-hubii-network-ico-floyd-crypto-mayweather-2017-8 (Accessed September 30, 2025)

[31] Scott, Allen. Mike Tyson Partners with Bitcoin Direct, LLC. September 15, 2015. https://cointelegraph.com/news/mike-tyson-partners-with-bitcoin-direct-llc (Accessed September 29, 2025)

[32] Sharma, Rakesh. Winklevoss Twins are Bitcoin's First Billionaires. December 4, 2017. https://www.investopedia.com/news/winklevoss-twins-are-bitcoins-first-billionaires/ (Accessed September 29, 2025)

[33] Santos, Maria. Sorry, Snoop Dogg! Mel B is the First Music Artist to Embrace Bitcoin. 99Bitcoins.com, January 2, 2018. https://99bitcoins.com/news/sorry-snoop-dogg-mel-b-is-the-first-music-artist-to-embrace-bitcoin/ (Accessed October 26, 2025)

- **Nas:** Rapper and technology investor Nas invested in numerous cryptocurrency startups, including Coinbase.[34]

Television/Film

- **Ashton Kutcher:** A movie, television star, and tech investor, invested in Unikrn, an e-sports betting digital platform with applications for skill-based gaming and virtual reality experiences. Unikrn developed its own cryptocurrency, UnikoinGold (UKG), which is used for transactions on the platform. The SEC later found that Unikrn did not register UKG as a security, and UKG agreed to pay a $6.1 million fine; as a result, UKG was disabled. [35]
- **Jamie Foxx:** Oscar-winner Jamie Foxx endorsed the ICO of Cobinhood Exchange to his five million 𝕏 followers. Since then, Cobinhood halted trading and allowed users to withdraw funds.[36]
- **Gwyneth Paltrow**, an Oscar, Golden Globe, and Primetime Emmy Award winner, Gwyneth Paltrow joined the Bitcoin-based start-up Abra as an Advisor. She pitched Abra during an episode of Planet of the Apps, a reality television show.[37]

---

[34] Sanchez, Melissa. "From Escobar to Investo-Bar: Investments by Nas Continue to Run Up His Net Worth." The Source, February 7, 2023. https://thesource.com/2023/02/07/from-escobar-to-investo-bar-investments-by-nas-continue-to-run-up-his-net-worth/ (Accessed September 29, 2025)

[35] Soper, Taylor. eSports Betting Startup Unikrn to Pay $6.1 Million in SEC Settlement Over Unregistered Initial Coin Offering. GeekWire.com, September 15, 2020. https://www.geekwire.com/2020/esports-betting-startup-unikrn-pay-6-1m-sec-settlement-unregistered-initial-coin-offering/#:~:text=Seattle%20startup%20Unikrn%20will%20pay%20a%20$6.1,offering%20in%20violation%20of%20federal%20securities%20laws. (Accessed October 28, 2025)

[36] Cheng, Yilung. Crypto Exchange Behind Jamie Foxx Endorsed ICO is Temporarily Shutting Down. TheBlock.com, January 10, 2020. https://www.theblock.co/linked/52637/crypto-exchange-behind-jamie-foxx-endorsed-ico-is-temporarily-shutting-down (Accessed October 28, 2025)

[37] Helms, Kevin. Gwyneth Paltrow Joins Bitcoin Wallet Abra as Advisor, Bitcoin.com, August 3, 2017. https://news.bitcoin.com/gwyneth-paltrow-joins-bitcoin-wallet-abra-as-advisor/ (Accessed October 28, 2025)

THE BIG PICTURE

The above examples demonstrate how cryptocurrencies are featured in mainstream media, from TV shows and movies to sporting events and social media. This clearly indicates their increased popularity and deep integration into pop and sporting cultures. By featuring them prominently in popular media, the industry demystifies cryptocurrencies, making them more understandable and appealing to the general public. Legions of fans follow celebrities and their actions, for better or for worse. Celebrity endorsement can lead to increased revenue streams, enhanced fan engagement, broader access to global audiences, and increased adoption and investment in cryptocurrencies. However, celebrity endorsements can also persuade uninformed individuals to invest in a fraudulent or non-compliant cryptocurrency. This is why cryptocurrency investors must conduct independent research before investing.

SUMMARY

This chapter underscores the importance of staying up to date on the rapidly evolving landscape of cryptocurrencies. One significant development discussed is the increasing integration of cryptocurrencies into popular culture over the last five years. From movies and TV shows to music and social media, cryptocurrencies are becoming more mainstream, attracting a wider audience and creating potential investors.

The chapter also explored the profound impact of the global pandemic on the cryptocurrency market. The combination of social isolation due to social distancing and economic uncertainty resulting from a major shift in business norms led to increased interest in cryptocurrencies as a potential hedge against inflation and traditional financial systems.

Additionally, the chapter explores the significance of Bitcoin and Ethereum spot ETFs and the impact of Bitcoin's halving. The approval of cryptocurrency spot ETFs in the U.S. could broaden the investor base for these digital assets. Also, Bitcoin halvings have historically led to price increases due to reduced supply.

Ultimately, this chapter highlights the dynamic nature of the cryptocurrency market and the importance of staying informed about emerging trends, new financial products, regulatory developments, and technological advancements. By understanding these factors and conducting thorough research, you'll be better equipped to navigate the complexities of investing in cryptocurrencies.

SUCCESS STRATEGY ACTION ITEMS

1. Follow official accounts, newsletters, and podcasts of the cryptocurrencies or blockchain projects you are interested in. They often release important news and updates about recent developments.
2. Follow reputable cryptocurrency news websites like CoinDesk, Cointelegraph, Decrypt, and The Block. These websites offer regular updates, market analyses, and insights into the cryptocurrency market.
3. Follow experienced and credible crypto experts, and official accounts on platforms like X, Reddit, and LinkedIn to find breaking news, analysis, and discussions on various cryptocurrencies.
4. Join crypto-related groups or channels where enthusiasts and experts discuss market trends, news, and share insights.
5. Attend crypto conferences, webinars, and events both physically and virtually. These gatherings often provide the latest industry trends, expert insights, and valuable networking opportunities.

CHAPTER 3

MAJOR CRYPTOCURRENCIES BY MARKET CAPITALIZATION

"The stock market is a device to transfer money from the impatient to the patient." [38]

Warren Buffett
Chairman and CEO, Berkshire Hathaway

CHAPTER OVERVIEW

This chapter focuses on the investment aspect of cryptocurrencies. Can you imagine that the first cryptocurrency was not intended as an investable asset at all? Emerging in response to the 2008 financial crisis, Bitcoin was designed to serve as an alternative to fiat currencies. Today, however, Bitcoin is used primarily for trading and investment purposes. Most people who buy Bitcoin do not use it to purchase every day or big-ticket items — they simply want their investment to increase in value.

Warren Buffett, one of the richest men on earth and among the greatest investors of all time, has offered excellent investment advice for decades. Although the above quote did not specifically refer to the cryptocurrency market, it remains applicable. I know people who lost money in the

[38] Singh, Navdeep. Five Timeless Warren Buffett Quotes Every Investor Should Know. The Economic Times, May 04, 2025. https://economictimes.indiatimes.com/markets/stocks/news/5-timeless-warren-buffett-quotes-every-investor-should-know/the-stock-market-is-a-device-for-transferring-money-from-the-impatient-to-the-patient-/slideshow/120868659.cms?from=mdr (Accessed October 28, 2025)

cryptocurrency market because they sold their cryptocurrency when the market was declining. Part of the reason for the transfer of wealth from impatient to patient investors is that patient investors hold onto their cryptocurrency and wait for the right time to sell at a higher price.

Investors use the acronym "HODL" to describe the patience of not selling a cryptocurrency in a bear market. HODL stands for "hold on for dear life" — I use HODL to say, "hold on and don't lose." Whatever meaning you choose to apply, HODL works best for investors with a long-term investment strategy. The cryptocurrency investment checklist is one strategy I found effective for selecting cryptocurrencies that meet my investment criteria. The Cryptocurrency Investment Checklist I designed for my cryptocurrency investment follows:

	(NAME OF CRYPTOCURRENCY)	Strongly Disagree -1	Disagree - 2	Agree - 3	Strongly Agree - 4	Neutral - 0
1.	You understand the company's business model.					
2.	The cryptocurrency adds value to business, industry, or society.					
3.	You were able to read a well-written white paper.					
4.	The company has been in existence for over five years.					
5.	The company has an experienced CEO and executive team.					
6.	The company submitted all of the required regulatory filings.					

	(NAME OF CRYPTOCURRENCY)	Strongly Disagree -1	Disagree - 2	Agree - 3	Strongly Agree - 4	Neutral - 0
7.	The company has a well-designed professional website.					
8.	The company has an active and responsive social media presence.					
9.	The company has a high market capitalization on many cryptocurrency exchanges.					
10.	At least three independent publications or organizations have published positive reviews or comments about the cryptocurrency.					
11.	Are there any other factors you believe are important to include in the checklist?					

It is essential to note that the numerical value assigned to each cell is based on the most current and accurate information available about the company. In addition, the "other factors" in row 11 of the checklist could include project developments, partnerships, regulatory news, and overall market sentiment. Determining other factors will require additional research; however, it is best to gather as much information as possible before investing in the cryptocurrency.

If you choose a cryptocurrency that does not meet the investment criteria, the virtue of patience might not be rewarded — no matter how long you HODL, you'll have a greater chance of losing some or all of your investment.

MARKET CAPITALIZATION RECAP FROM *BLOCKCHAIN OR DIE*

This chapter provides an update to Chapter 2, "Top Ten Cryptocurrencies by Market Capitalization," in *Blockchain or Die.* Most of the top 10 cryptocurrencies by market capitalization have remained in the top 10 for several years, suggesting "blue chip" characteristics of longevity and consistency. Many of these cryptocurrencies are arguably less risky investments, as they have consistently retained their market capitalization for several years.

As a reminder, here are the top ten cryptocurrencies by market capitalization, as featured in *Blockchain or Die.*

On June 12, 2018, the top ten cryptocurrencies by market capitalization were:

#	Name	Market Cap	Price
1.	Bitcoin	$70,762,404,175	$4,050.43
2.	Ethereum	$15,746,707,219	$151.03
3.	XRP (Ripple)	$14,898,900,670	$0.365222
4.	Bitcoin Cash	$2,825,737,651	$160.96
5.	EOS	$2,545,625,158	$2.81
6.	Litecoin	$2,389,652,158	$39.87
7.	Stellar	$2,362,374,602	$0.123276
8.	Tether	$1,906,175,130	$1.02
9.	TRON	$1,738,132,471	$0.026081
10.	Bitcoin SV	$1,546,736,488	$88.11

On August 28, 2018, the top ten cryptocurrencies by market capitalization were:

#	Name	Market Cap	Price
1.	Bitcoin	$144,683,199,615	$8,148.65
2.	Ethereum	$27,396,567,735	$257.33

#	Name	Market Cap	Price
3.	XRP (Ripple)	$17,027,601,016	$0.440361
4.	Litecoin	$8,342,138,256	$134.13
5.	Bitcoin Cash	$7,035,302,870	$398.48
6.	EOS	$5,876,766,250	$6.40
7.	Binance Coin	$4,881,874,908	$34.58
8.	Bitcoin SV	$3,385,063,692	$189.83
9.	Tether	$3,373,283,586	$1.01
10.	Stellar	$2,452,158,584	$0.126404

When *Blockchain or Die* was published, CoinMarketCap listed 2,235 cryptocurrencies. The 2018 tables showed that Bitcoin, Ethereum, and Ripple retained their respective top 1–3 positions in market capitalization between June 2018 and August 2018. The 4–10 cryptocurrencies showed some changes in position, but over this period, as quoted in the chart, Bitcoin, Ethereum, XRP (Ripple), Bitcoin Cash, EOS, Stellar, Litecoin, and Tether have consistently remained in the top ten.[39]

Seven years later, how have the top cryptocurrencies by market capitalization (market cap) changed? How many cryptocurrencies are on CoinMarketCap.com?

As of August 2025, the total number of cryptocurrencies increased from 2235 to over 9,500. Here are the top ten cryptocurrencies by market capitalization.

#	Name	Market Cap	Price
1.	Bitcoin	$2,237,261,226,131	$112,360.
2.	Ethereum	$552,533,178,682	$4,577
3.	XRP	$175,687,693,341	$2.95
4.	Tether	$167,122,253,183	$0.9998

[39] "Top 100 Cryptocurrencies," *CoinMarketCap.com*. https://coinmarketcap.com (Accessed September 3, 2025)

#	Name	Market Cap	Price
5.	BNB	$120,494,670,075	$865
6.	Solana	$106,326,853,519	$196
7.	USDC	$67,532,225,720	$0.9998
8.	TRON	$33,223,577,165	$0.3509
9.	Dogecoin	$32,816,014,673	$0.2178
10.	Cardano	$31,086,662,989	$0.8701

In my opinion, the top ten cryptocurrencies by market capitalization are important as they represent the most established and widely adopted assets in the market. They typically signify higher liquidity and greater investor confidence, often serving as benchmarks for the overall health and direction of the cryptocurrency market. This list serves as a benchmark that every cryptocurrency investor should track regularly.

The upcoming section will examine the top ten cryptocurrencies by market capitalization as of the publication date.

ANALYSIS OF TOP TEN CRYPTOCURRENCIES BY MARKET CAPITALIZATION

The CoinMarketCap.com market capitalization section provides an overview of the value of cryptocurrencies in circulation, offering an overall valuation. Market capitalization is one of several factors that help investors make informed decisions about the quality of cryptocurrency investments.

The analysis of the top ten cryptocurrencies by market capitalization below includes the following information:

- **CoinMarketCap.com information:** Name, price, chart, etc.
- **Currency Name and Symbol:** Some cryptocurrency businesses name their currency after their company, while others give it a distinct name. The analysis includes the symbol to help you identify the cryptocurrency quotes.

- **Year Created:** Shows the longevity of the cryptocurrency.
- **Creator(s):** The success of the business is often based on the experience and track record of the founder or CEO. Knowing the creator's name facilitates further research.
- **Industry:** Identifies the specific sector the cryptocurrency is designed to serve or to evolve into.
- **General Information:** Information about each cryptocurrency is directly quoted from its website.
- All of the graphs were downloaded in January 2026, from CoinMarketCap.com. For each cryptocurrency, the first graph displays the entire price history of the asset. The second graph shows the price history of the cryptocurrency from January 2022.

#1 BITCOIN

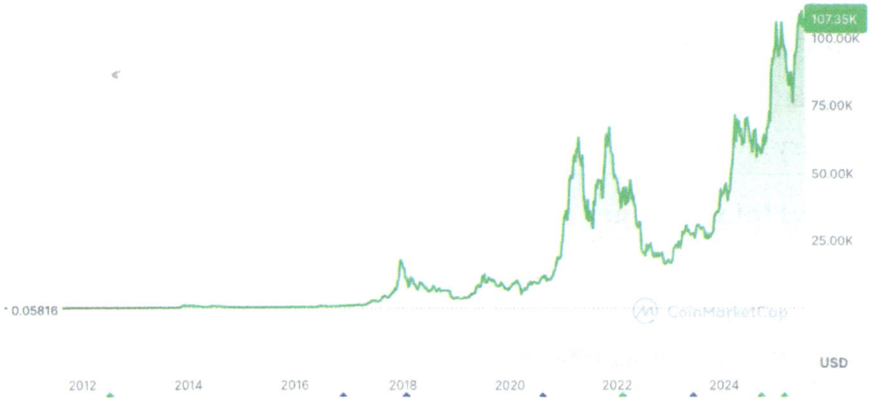

Chart #2, starting in January 2022, shows that Bitcoin's value increased significantly, consistently breaking new price barriers and reaching $107,000 by June 2025. While there were corrections and periods of consolidation, the overall trend was a substantial increase.

Currency Name: Bitcoin (BTC)
Year Created: 2009
Creator(s): Satoshi Nakamoto
Primary Industries: Currency

General Information:

The Bitcoin summary on Bitcoin.com remains unchanged from its 2019 version.

> Bitcoin uses peer-to-peer technology to operate without central authorities or banks; managing transactions and issuing of bitcoin is carried out collectively by the network. Bitcoin is an open-source cryptocurrency; its design is publicly available, and nobody owns or controls it, allowing everyone to participate. Through many of its unique properties, Bitcoin enables exciting uses that cannot be covered

by any previous payment system.[40]

The following diagram gives a fundamental description of the Bitcoin transaction process:

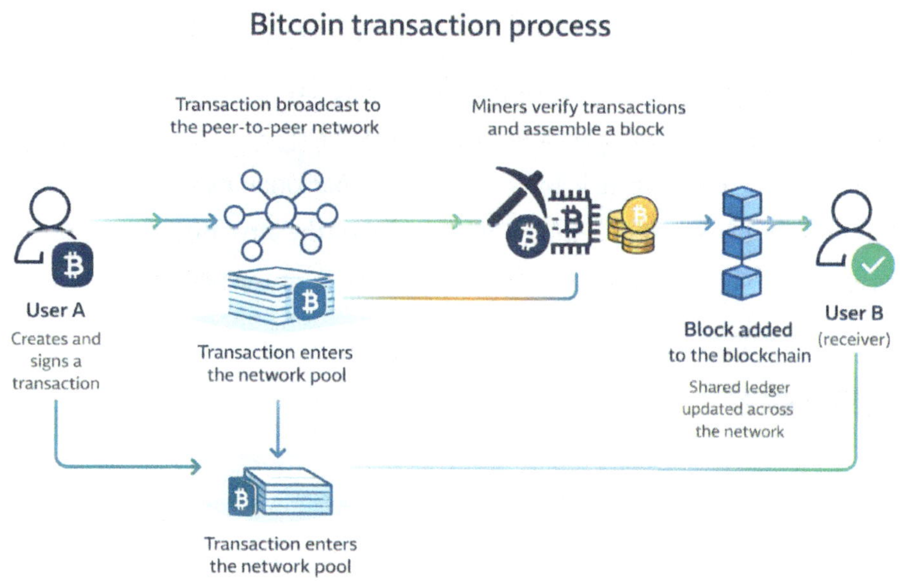

Although this is the same Bitcoin summary from 2019, it does not reflect the updates to the Bitcoin protocol, such as the Taproot upgrade covered in the previous chapter. Bitcoin, which has become a global financial and cultural phenomenon, has evolved significantly since its inception; the original white paper seems like ancient history. In addition, there are over 70 active forks of the Bitcoin network as of November 2023. This number has fluctuated over time, with some forks becoming inactive or merging with others.[41]

---

[40] Bitcoin. https://bitcoin.org/en/ (Accessed October 28, 2025)

[41] Weston, Georgia. A Complete List of Bitcoin Forks. 101Blockchains.com, August 23, 2024. https://101blockchains.com/bitcoin-forks-list/#:~:text=As%20of%20November%202023%2C%20there,forks%20of%20the%20Bitcoin%20network. (Accessed October 28, 2025)

With so much happening in the Bitcoin universe, Bitcoin.com is the premier destination for the latest news on this cryptocurrency. The Bitcoin White Paper is a great place to start your journey, and when you are ready to bring your Bitcoin journey to the present day, visit Bitcoin.com.

Later, Satoshi Nakamoto withdrew from the project, leaving it in the hands of a small but devoted group of developers. His last known emails went to Gavin Andresen, who became Bitcoin's de facto lead. The handoff was critical as it transformed Bitcoin from a personal experiment into a community project. Hal Finney, who had run the first Bitcoin client and received the very first transaction from Satoshi, kept the network alive in its fragile early days.

#2 ETHEREUM

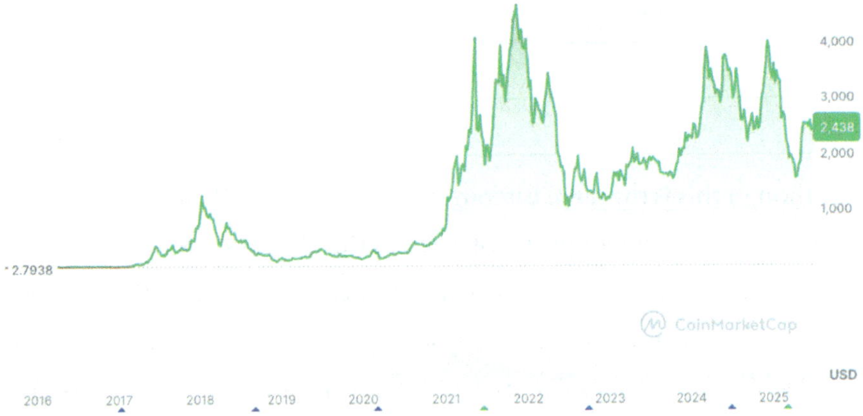

Chart #2, starting in January 2022, shows a significant drop in early 2024, followed by a strong recovery in April 2024. Ethereum's price increase reflects renewed optimism fueled by critical technological upgrades (especially the Merge and scaling solutions), increased institutional interest, and the continued growth of its DeFi and NFT ecosystems.

Currency Name: Ether (ETH)
Year Created: 2015
Creator(s): Vitalik Buterin
Primary Industries: Application Platform

General Information:

> Ethereum is a technology that powers digital money, global payments, and applications. The community has built a thriving digital economy, offering bold new ways for creators to earn online, and much more. It's open to everyone, wherever you are in the world – all you need is an internet connection.[42]

The words in the quote above that I find most impactful are: "the community has built a booming digital economy," because they are

[42] Ethereum. www.Ethereum.org https://ethereum.org/en/. (Accessed October 28, 2025)

absolutely true. In my opinion, cryptocurrencies with a solid business case and value proposition experienced significant growth after Bitcoin, largely due to Ethereum's evolution of the Bitcoin blockchain into the business space. The creation of smart contracts, dApps, and decentralized autonomous organizations is mostly based on Ethereum's vision of building a booming digital economy. Ethereum's impact on the digital community is also evident in the fact that numerous blockchain companies have built their protocols on Ethereum's business blockchain model.

#3 TETHER

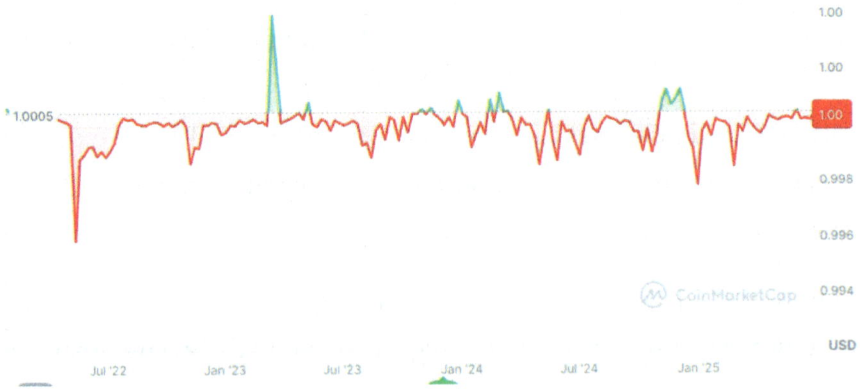

Chart #2 shows that Tether's performance is measured by its stability, rather than its appreciation. Tether's price performance between January 2022 and June 2025 is primarily characterized by its successful adherence to its US Dollar peg. The most significant event was the de-pegging in May 2022, which resulted in a temporary price drop. However, its swift recovery and consistent performance since then have reinforced its role as a reliable stablecoin.

Currency Name: Tether (USDT)
Year Created: 2012
Creator(s): Craig Sellers, Reve Collins, and Brock Pierce

General Information:

> Tether is a token backed by actual assets, including the US dollar and the euro. One Tether equals one underlying unit of the currency backing it, e.g., the U.S. Dollar, and is supported 100% by actual assets in the Tether platform's reserve account. Being anchored or "tethered" to real-world currency, Tether provides protection from the volatility of cryptocurrencies. Tether enables businesses—including exchanges, wallets, payment processors, financial services, and ATMs— to easily use fiat-backed tokens on blockchains."[43]

Many cryptocurrency traders I know use Tether to trade and transfer value between cryptocurrency investments, thereby avoiding the need to convert their cryptocurrencies into fiat currencies. The fact that Tether is backed by assets gives it the stability and comfort to "park" its cryptocurrencies without risking loss in a high-risk investment. For this reason, Tether is more akin to an actual currency than Bitcoin, which is more commonly used as an investment or commodity.

---

[43] Tether. www.Tether.to https://tether.to/en/. (Accessed October 28, 2025)

Since Tether is a stablecoin, its chart looks very different. Aside from a few isolated upward and downward spikes, Tether usually trades very close to 1.00 USD per coin.

Finally, it is worth noting that while Bitcoin is the #1 cryptocurrency by market capitalization, Tether is the #1 cryptocurrency by volume.

#4 XRP

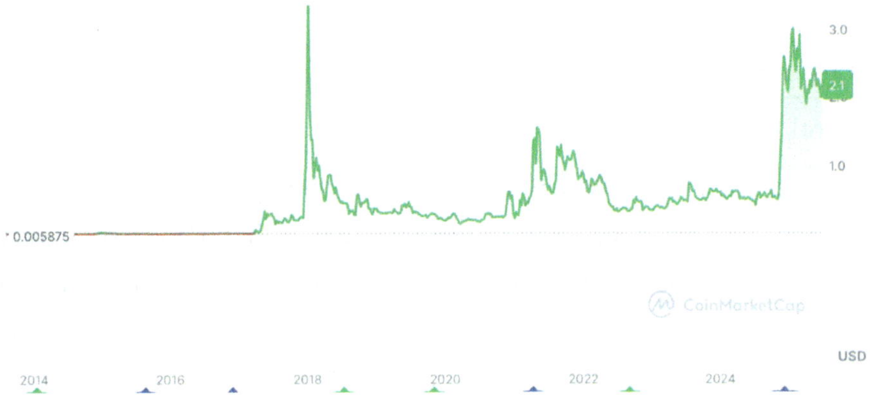

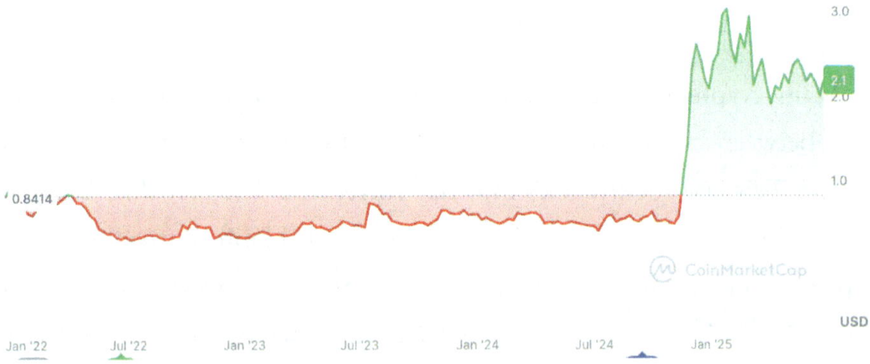

Chart #2 shows a significant price increase for XRP. Several events contribute to the price increase. Between January 2022 and June 2025, XRP navigated a challenging regulatory environment and broader market downturns. However, the partial resolution of the SEC lawsuit in mid-2023 and the full resolution in 2025 contributed to the price increase. In addition, XRP benefited from growing market optimism and the return of institutional investors.

Currency Name: XRP
Year Created: 2016
Creator(s): Chris Larsen and Jeb McCaleb
Primary Industries: Payment and Exchange Network

General Information:

> Through blockchain technology, Ripple enables global financial institutions, businesses, governments, and developers to move, manage, and tokenize value, unlocking greater economic opportunities for everyone, everywhere. [44]

When I first read about Ripple, I noticed two things: first, Ripple was centralized, unlike other cryptocurrencies, and second, Ripple's business model posed a significant threat to money transmission services (e.g., Western Union) and international banks (such as JPMorgan Chase and Bank of America). It makes sense that Ripple is centralized, since its primary clients are financial institutions.

In fact, Ripple makes that point clear in their white paper: "Ripple connects banks, payment providers, digital asset exchanges, and corporates via Ripplenet to provide one frictionless experience to send money globally." The threat that Ripple's business model poses to traditional money transfer services and banks has not gone unnoticed; some of these institutions have

---

[44] About Ripple. Ripple.com. https://ripple.com/company/ (Accessed September 3, 2025)

even partnered with Ripple to transmit XRP globally, including Santander, American Express, and UBS.[45]

#5 BNB

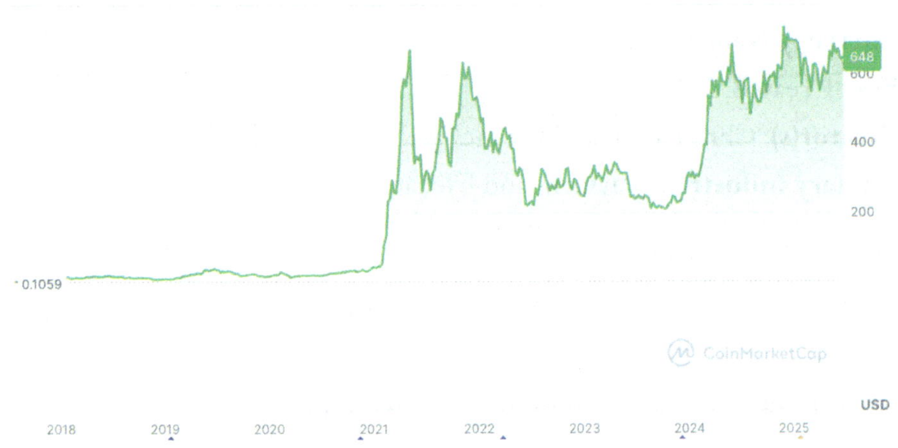

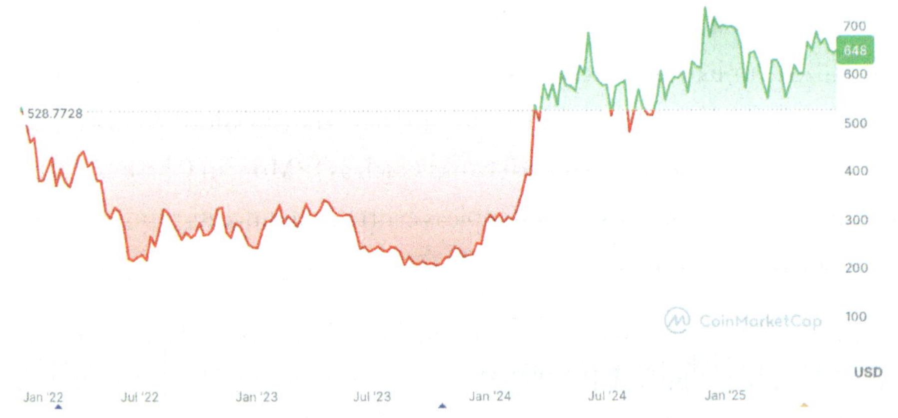

Chart #2 shows a significant price increase for BNB. The growth of the BNB ecosystem provided underlying support. In addition, in May 2025, the SEC dismissed its lawsuit against Binance. This removed a major regulatory roadblock, boosting confidence in Binance and, by extension, BNB.

[45] B, Andrew. Six Biggest Banks Using Ripple (XRP) Products. UseTheBitcoin.com, January 2, 2025. https://usethebitcoin.com/resources/6-biggest-banks-using-ripple-products/ (Accessed October 28, 2025)

Currency Name: Binance Coin (BNB)
Year Created: 2017
Creator(s): Changpeng Zuao and Yi He
Primary Industries: Cryptocurrency Exchange

General Information:

Binance is a blockchain and cryptocurrency exchange that provides the following services:

- Academy—Blockchain and crypto-education
- BCF—Blockchain charity foundation
- Info—Cryptocurrency information platform
- Labs—Incubator for top blockchain projects
- Launchpad—Token launch platform
- Research—Institutional-grade analysis and reports
- Trust Wallet—Binance's official crypto wallet. [46]

Additionally, BNB is a platform token and the native token of the Binance Smart Chain (BSC), a blockchain specifically designed for smart contracts and decentralized applications (dApps). This integration fuels the demand for BNB for gas fees on the BSC network.

---

[46] Binance. https://www.binance.com/en/about. (Accessed September 3, 2025)

#6 SOLANA

Chart #2 shows a significant price increase for SOL. This increase was due to upgrades to the Solana blockchain and the return of institutional investors.

Currency Name: Solana (SOL)
Year Created: 2009
Creator(s): Anatoly Yakoenko
Primary Industries: Currency

General Information:

Solana is a highly functional open-source project that banks on the permissionless nature of blockchain technology to provide decentralized finance (DeFi) solutions. While the idea and initial work on the project began in 2017, Solana was officially launched in March 2020 by the Solana Foundation, headquartered in Geneva, Switzerland.[47]

The Solana protocol is designed to facilitate the creation of decentralized apps (DApps). It improves scalability by introducing a proof-of-history (PoH) consensus combined with the blockchain's underlying proof-of-stake (PoS) consensus.

Due to its innovative hybrid consensus model, Solana attracts interest from both small-time traders and institutional investors. The Solana Foundation places significant focus on making decentralized finance accessible on a larger scale.

Solana is another platform token and is unique in that it uses a Proof of History (PoH) consensus mechanism to timestamp transactions. Unlike traditional consensus mechanisms such as Proof of Work (PoW) or Proof of Stake (PoS), which rely on network consensus to determine the order of events, PoH provides each node with a local clock. Solana can handle thousands of transactions per second (TPS), significantly more than many other blockchains, such as ETH and Bitcoin.

Solana also has the Solana Foundation, which provides grants to projects that generate public goods for the community. The Solana Foundation has distributed over $100 million in grants. The Solana Foundation may even help you fund your project if it generates goods for the community.

---

[47] Solana, Flagship. https://flagship.fyi/glossary/solana-sol/ (Accessed October 31, 2025)

#7 USDC

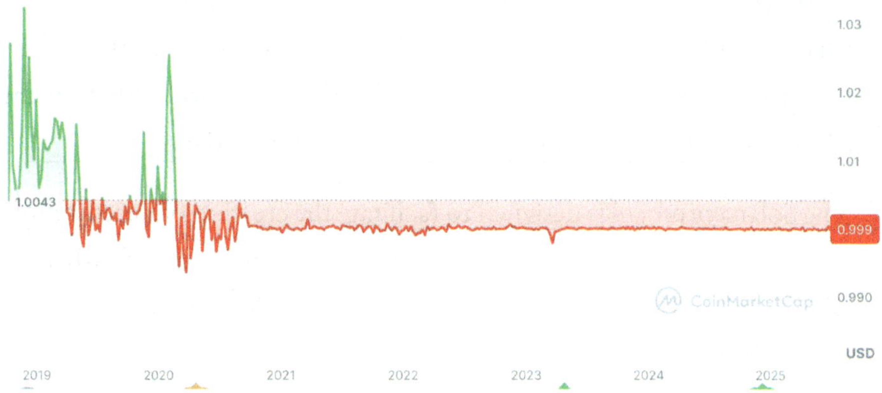

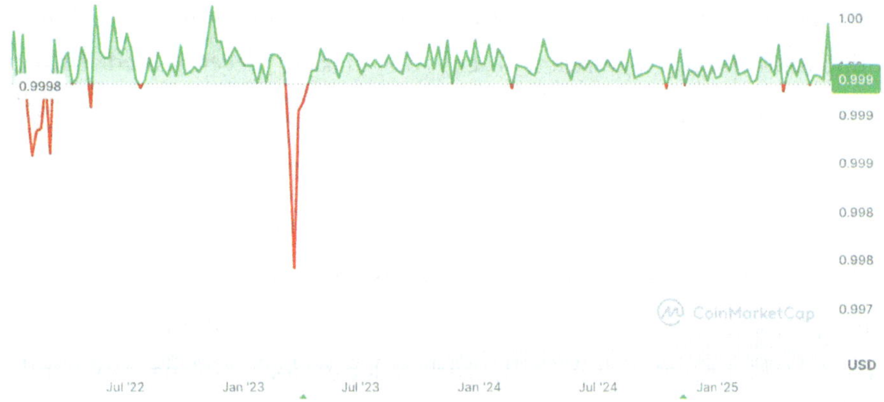

Chart #2 shows that USDC's price performance from January 2022 to June 2025 has been characterized by consistent adherence to its dollar peg, with the notable exception of a temporary de-peg during the Silicon Valley Bank crisis, when US Dollars tied to USDC were held. USDC's stability is attributed to its strong reserve backing, transparent auditing, proactive regulatory compliance, and growing utility within the decentralized finance and payment ecosystems.

Currency Name: USD Coin
Year Created: 2013

Creator(s): The Centre Consortium (Circle and Coinbase)
Primary Industries: Currency

General Information:

USDC is the world's largest, regulated, fully reserved digital dollar. It's backed 100% by highly liquid cash and cash-equivalent assets and is redeemable at a 1:1 ratio for US dollars. Leveraging the power of the internet and blockchain technology, UDC can move across borders quickly with low fees and is accessible to individuals with an internet connection in more than 180 countries worldwide.[48]

Like Tether, USDC is a stablecoin, but with one main difference: it is pegged to the US dollar, meaning its value is designed to remain closely tied to the US dollar. This is achieved by backing each USDC token with a US dollar reserve.

I have noticed that Circle and Coinbase are very transparent about USDC. The Circle.com website points out that the majority of the USDC reserve is invested in the Circle Reserve Fund (USDXX), a SEC-registered 2a-7 government money market fund. The Circle.com website connects viewers to BlackRock, an independent third-party reporting service for the Circle Reserve Client fund.[49]

---

[48] Foundations. What is USDC? USDC.com. https://www.usdc.com/learn/what-is-usdc (Accessed October 27, 2025)

[49] Circle Reserve Fund, Blackrock.com, October 24, 2025. https://www.blackrock.com/cash/en-us/products/329365/ (Accessed October 27, 2025)

#8 TRON

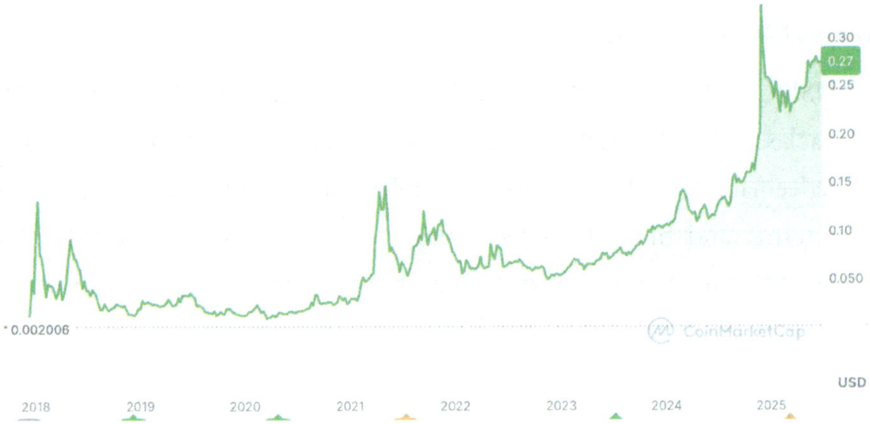

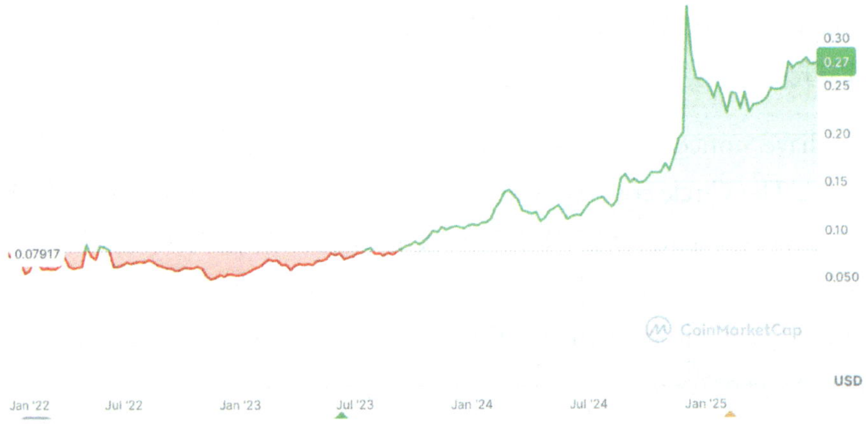

The second chart shows that TRON's price performance from January 2022 to June 2025 recovered from the 2022 bear market, followed by a strong bullish run in 2023 and 2024, and an overall sustained price increase in early to mid-2025. This performance has been largely driven by TRON's robust ecosystem and increasing institutional interest.

Currency Name: Tron (TRX)
Year Created: 2017
Creator(s): Justin Sun

Primary Industries: Currency

General Information:

TRON (TRX) is a decentralized blockchain-based operating system developed by the Tron Foundation and launched in 2017. Originally, TRX tokens were ERC-20-based tokens deployed on the Ethereum network, but a year later, they were migrated to their own blockchain.

The project was initially created to provide full ownership rights to makers of digital content. The primary goal is to support content creators (who typically receive only a small portion of the income) and reward them with additional incentives for their work.

The TRON software supports smart contracts, various blockchain systems, and decentralized applications (dApps). Transactions are recorded in a public ledger, allowing users to track their history.

TRON became very popular among gamers, likely due to its namesake, the groundbreaking movie "TRON," which introduced the world to digital gaming. Moreover, as a decentralized blockchain, TRON has an active community that supports its continued development.

#9 DOGECOIN

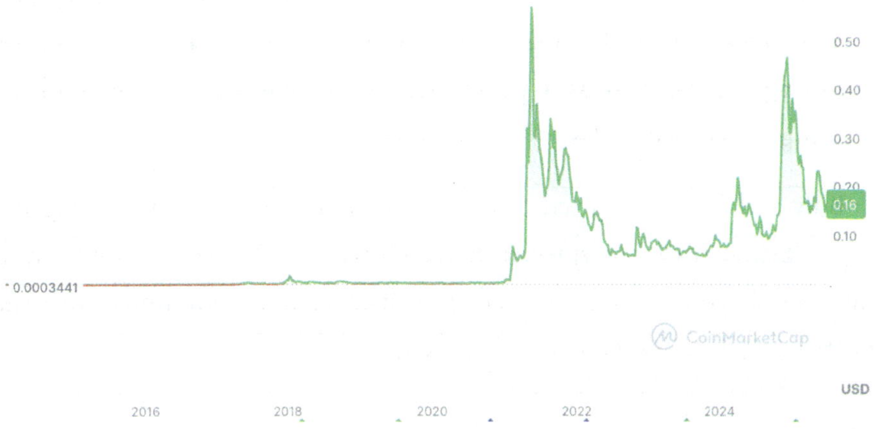

The second chart shows DOGE's volatility as a meme coin. DOGE barely recovered in 2024 before another price drop, then experienced a significant price increase in 2025. The 2025 price increase was short-lived, as DOGE lost all its gains in a short period.

Currency Name: Dogecoin (DOGE)
Year Created: Forked from Litecoin in 2013
Creator(s): Billy Markus and Jackson Palmer
Primary Industries: Currency

General Information:

At its heart, Dogecoin is the accidental crypto movement that makes people smile! It is also an open-source peer-to-peer cryptocurrency that uses blockchain technology, a highly secure, decentralized system for storing information as a public ledger maintained by a network of computers called nodes.

Billy and Jackson started Dogecoin to satirize Bitcoin by turning the "doge" meme into a cryptocurrency. However, Dogecoin evolved into something beyond a cryptocurrency — it created a community with the unofficial tagline: "*Do Only Good Every day.*"

There is another major difference between Doge and Bitcoin, as well as other cryptocurrencies in the CoinMarketCap Top Ten: Doge has a very high total supply of 145.8 billion tokens, compared to Bitcoin's total supply of 21 million. In a supply-and-demand analysis, Dogecoin's high supply may always lead to lower demand, resulting in a lower price.[50]

#10 CARDANO

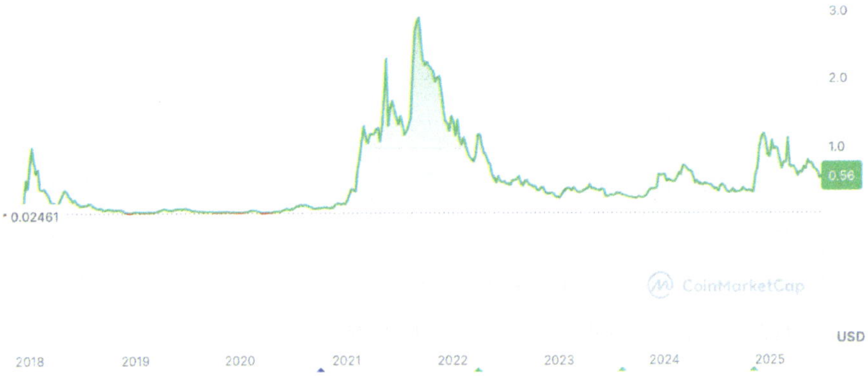

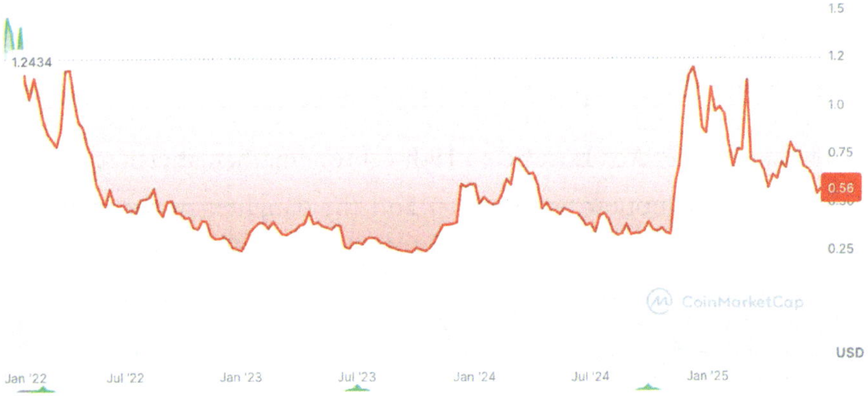

[50] Dogecoin. CoinMarketCap.com. https://coinmarketcap.com/currencies/dogecoin/ (Accessed September 3, 2025)

The second chart shows that Cardano experienced significant price volatility. Cardano experienced a sharp decline in 2022. However, it demonstrated a notable recovery in 2024 and 2025, driven by its ongoing technological advancements and growing ecosystem.

Currency Name: Cardano (ADA)
Year Created: 2017
Creator(s): Charles Hoskinson (co-founder of Ethereum)
Primary Industries: Currency

General Information:

According to Cardano.org:

> Cardano is a proof-of-stake blockchain platform: the first to be founded on peer-reviewed research and developed through evidence-based methods. It combines pioneering technologies to provide unparalleled security and sustainability to decentralized applications, systems, and societies.
>
> With a leading team of engineers, Cardano exists to redistribute power from unaccountable structures to the margins – to individuals – and be an enabling force for positive change and progress.[51]
>
> It is named after Ada Lovelace, a 19th-century mathematician recognized as the first computer programmer and the daughter of the poet Lord Byron.[52]

Many people in the blockchain space view Cardano as a socially responsible blockchain company. Cardano leadership paid homage to Ada Lovelace by naming their token "ADA. Moreover, Cardano features Project

---

[51] Cardano. https://cardano.org (Accessed September 30, 2025)

[52] What is ADA? https://cardano.org/what-is-ada (Accessed September 30, 2025)

Catalyst, an ambitious initiative designed to harness decentralized innovation and address real-world challenges. Project Catalyst has already distributed over $100 million to blockchain projects. If your cryptocurrency solution has the potential to solve real-world challenges, Project Catalyst might help fund your project!

THE BIG PICTURE

Despite the emergence of 9,400 altcoins, Bitcoin, Ethereum, and Ripple continue to hold the top three positions by market capitalization, reflecting their strong foundations and widespread adoption. Four of the top ten cryptocurrencies by market capitalization are platform tokens (ETH, BNB, ADA, and SOL), underscoring the significance of blockchain technology as a foundation for various applications. Two of the tokens, USDT and USDC, are stablecoins, which demonstrates the widespread use of stablecoins in the cryptocurrency market. Finally, tokens that leverage the power of community involvement, such as ADA, DOGE, and TRON, demonstrate the importance of community in driving the success of cryptocurrencies.

COINMARKETCAP FEAR AND GREED INDEX

Whenever I think about greed and investing, I think of the movie Wall Street (1987) and the speech given by Michael Douglas's character, Gordon Gekko, about greed and investing.

> [G]reed -- for lack of a better word -- is good. Greed is right. Greed works. Greed clarifies, cuts through, and captures the essence of the evolutionary spirit. Greed, in all its forms — greed for life, for money, for love, and for knowledge—has marked the upward surge of mankind.[53]

[53] Wall Street, directed by Oliver Stone (1987; 20th Century Fox).

Contrary to what moral fables and stories may suggest, greed in investing can be beneficial, provided you combine it with thorough research and smart, patient investing strategies, such as waiting for the market to display bullish signs. These signs can be found in the CoinMarketCap Fear and Greed index.

The CoinMarketCap Fear and Greed Index, launched in July 2023, is an online tool that gauges the sentiment of the cryptocurrency market, offering a score from 0 to 120. The index reflects investors' sentiment towards the market at a given time, indicating potential buying or selling opportunities. The index further uses the following data points to capture the overall market:

1. **Price Momentum:** This factor analyzes the price performance of the top ten cryptocurrencies by market capitalization (excluding stablecoins). It assesses how these coins are performing relative to each other and the broader market.
2. **Volatility:** The index incorporates Volmex Implied Volatility Indices (BVIV and EVIV) for Bitcoin (BTC) and Ethereum (ETH), which provide forward-looking measures of expected volatility over the next 30 days.
3. **Derivatives Market:** The index considers the Put/Call Ratio in the Bitcoin and Ethereum options markets. A higher ratio of puts to calls indicates more fear in the market, suggesting bearish expectations among investors.
4. **Market Composition:** This component examines the relative value of Bitcoin (BTC) in the market, using the Stablecoin Supply Ratio (SSR) to compare Bitcoin's market capitalization with that of major stablecoins.

5. **CMC Proprietary Data:** The index also uses social trend keyword searches and user engagement metrics to capture market sentiment, retail interest, and emerging trends.[54]

As displayed below, the Fear and Greed Index (The Index) is on a scale between 0 (Extreme Fear) and 120 (Extreme Greed):

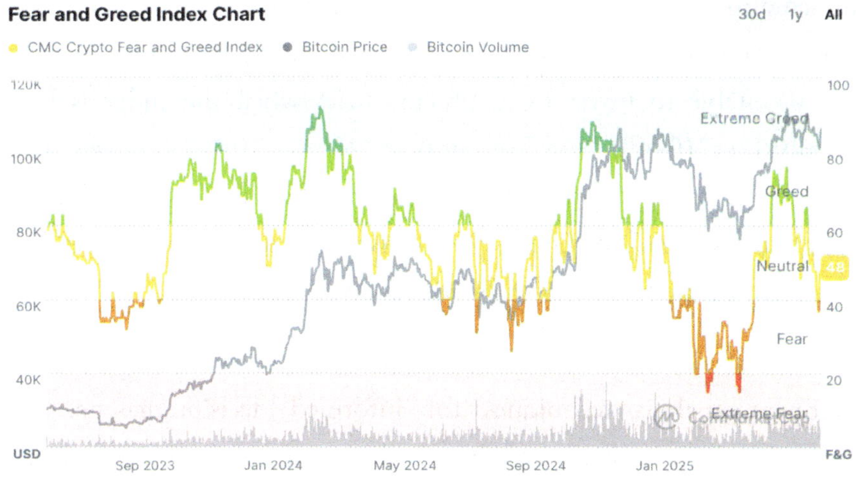

- **$0-40K: Extreme Fear** - This indicates that investors are extremely fearful, which may trigger a significant sell-off. However, this sell-off can create a buying opportunity for those with a long-term view.
- **$41K-60K: Fear** - Investors are cautious, but a potential buying opportunity might exist for risk-tolerant individuals.
- **$61K – 80K - Neutral** - The market sentiment is balanced between fear and greed.
- **$81K-100K - Greed** - Investors are becoming increasingly optimistic, and prices may be on the rise, but this could lead to market bubbles, setting the stage for a crash.

---

[54] CoinMarketCap Fear and Greed Index. https://coinmarketcap.com/charts/fear-and-greed-index/ (Accessed September 29, 2025)

- **101K-120K: Extreme Greed** - This indicates excessive optimism and euphoria, potentially signaling an overheated market that may be due for a correction.

IMPORTANT FEAR AND GREED INDEX CONSIDERATIONS

It is important to note that the Fear and Greed Index is a simplified representation of market sentiment and should be one of many factors when making investment decisions. From a technical trading standpoint, it may be more profitable to invest in cryptocurrencies when the Index is in the "Extreme Fear" (0–40), "Fear" (41–60), or "Neutral" (61–80) ranges. This is because, historically, cryptocurrencies as a whole have increased in value from "Fear" and "Neutral" to "Greed". Of course, this technical analysis does not account for the performance of individual cryptocurrencies.

The Fear and Greed Index should be used in conjunction with other investment analysis techniques for informed decision-making. As I mentioned earlier in these pages (and will continue to do so), it's crucial to conduct thorough research on specific cryptocurrencies before investing. Take a look at the project goals, technology, and the team behind them.

It is also important to note that the Fear and Greed Index does not always follow the same trajectory as Bitcoin. Notice that in January 2024 and January 2025, there is a large gap between Bitcoin and the rest of the cryptocurrencies in the Fear and Greed Index. This represents a substantial deviation, highlighting the significance of the Fear and Greed Index.

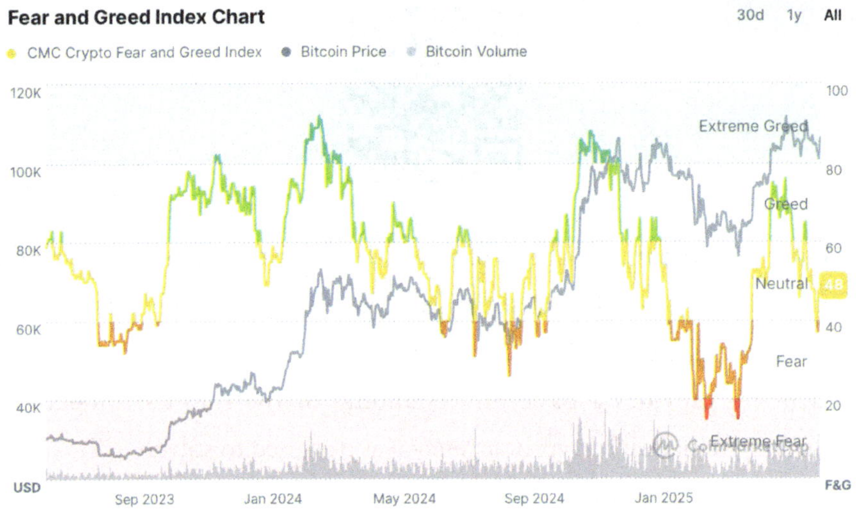

SUMMARY

As noted in this chapter, cryptocurrencies have advanced significantly since 2019 but remain in a transitional state, facing numerous challenges. However, despite the challenges, since August 2018, 4 out of 10 cryptocurrencies have remained in the top 10 by market capitalization on CoinMarketCap.com, indicating stability and longevity in the market. For their price, these cryptocurrencies have shown great promise through the products and services they offer.

SUCCESS STRATEGY ACTION ITEMS

1. The CoinMarketCap.com cryptocurrency graphs are an important feature that tells a cryptocurrency's trading history. If you are a cryptocurrency investor interested in tracking market performance, check CoinMarketCap.com or other cryptocurrency resources to identify cryptocurrencies that are outperforming other markets.

2. Visit the websites of the 2025 top ten cryptocurrencies by market capitalization to learn some of the best business practices in cryptocurrencies. While each website is distinct in format and content, each site provides a basic understanding of the business models and challenges that each cryptocurrency was designed to solve.
3. Use the Fear and Greed Index to gauge market sentiment when investing in cryptocurrencies. Don't forget that the Index is only one of the many investment tools you should use to make investment decisions.
4. Cardano Foundation's Project Catalyst is a potential funding source for blockchain startups. Anyone with an idea can submit a proposal on the Project Catalyst platform. Visit the Project Catalyst website to learn more about their grant and funding opportunities.

5. The Solana Foundation offers several grants and funding opportunities for blockchain companies, particularly start-ups. Visit the Solana website to learn more about their grant and funding opportunities.

CHAPTER 4

REGULATORY LANDSCAPE FOR CRYPTOCURRENCIES

The SEC "should be there with its ear to the ground to… accommodate activity that's not criminal and then enable markets to flourish, because if it challenges incumbents and it helps to bring down costs for investors and for people who are trying to raise capital, I mean that's the reason why we have financial markets". [55]

Paul Atkins
Chairperson, Securities and Exchange Commission

CHAPTER OVERVIEW

The cryptocurrency regulatory landscape is complex and rapidly evolving.

While the potential benefits of cryptocurrencies are increasingly recognized, concerns about consumer protection, market manipulation, and money laundering persist. Regulatory bodies, including the Securities and Exchange Commission, the Commodity Futures Trading Commission, the Department of the Treasury, and the Internal Revenue Service, are focusing on areas such as investor protection, anti-money laundering (AML) compliance, and tax reporting.

Paul Atkins' quote, advocating for the SEC to "accommodate activity that's not criminal and then enable markets to flourish" and "bring down costs for investors," holds significant implications for individual cryptocurrency investors. It suggests a regulatory philosophy that prioritizes

[55] The Banker. Trump Taps Crypto Advocate Paul Atkins as SEC Chair. The Banker.com, December 5, 2024. https://www.thebanker.com/content/c9678c62-0987-5108-8935-acbe675a8f24 (Accessed September 4, 2025)

fostering innovation and reducing barriers, rather than solely focusing on enforcement. For individual crypto investors, this outlook could lead to a more predictable and less volatile market environment, thanks to clearer regulations, which may reduce the risk of sudden market shifts driven by regulatory crackdowns. Moreover, if the SEC aims to "bring down costs," it could translate into lower trading fees, more accessible investment products (like spot Bitcoin ETFs), and potentially more transparent reporting from cryptocurrency platforms. This approach might encourage more mainstream adoption of crypto, further legitimizing it as an asset class and potentially increasing liquidity and stability for individual investors.

The majority of this chapter seeks to answer a fundamental question: *How does the government classify and regulate cryptocurrencies?*

Four different U.S. Government Agencies regulate cryptocurrencies:

1. **A commodity**—The Commodities and Futures Trading Commission[56]
2. **A security** —The Securities and Exchange Commission
3. **A currency**—Department of the Treasury, Financial Crimes Enforcement Network, and the U.S. Office of Foreign Assets Control
4. **Property**—The Internal Revenue Service

The fast-paced nature of the crypto industry often outpaces regulatory efforts. This has led to regulatory arbitrage, with crypto firms seeking jurisdictions with more lenient rules. As the industry matures and its impact on the global financial system becomes clearer, a more unified regulatory approach will likely emerge. Ultimately, most policymakers aim to strike a balance between fostering innovation and protecting investors.

---

[56] U.S. Commodities and Futures Trading Commission, "CFTC Statement on Self-Certification of Bitcoin Products by CME, CFE and Cantor Exchange, *U.S. Commodities and Futures Trading Commission.* December 1, 2017. https://www.cftc.gov/PressRoom/PressReleases/pr7654-17 (Accessed September 29, 2025)

SMART INVESTORS FOLLOW INSTITUTIONAL INVESTORS

As a former Licensed Series 7 Financial Advisor, I gained valuable investment insights that I'm excited to share with you. One key lesson is that smart individual investors follow the lead of institutional investors. Why? It's the movement of institutional money that really drives the market.

Institutional investors invest large sums of money in the cryptocurrency market and have dedicated departments and teams of employees researching cryptocurrency companies to determine if the investment risk is worth the potential reward. These companies thoroughly research issues such as levels of compliance before investing in a cryptocurrency. If the cryptocurrency company is not in compliance or the compliance risk is too high, institutional investors will likely refrain from investing.

As an individual investor, you do not have departments and teams of employees to spend hours researching cryptocurrency companies — you only have yourself and whatever time you can take out of your day to conduct compliance research on a cryptocurrency company. That is, if you even know how. Institutional investors have teams of analysts constantly conducting regulatory due diligence on cryptocurrency companies. If they find a regulatory issue or a lawsuit, they may decide not to invest in the cryptocurrency or sell it. These decisions may leave individual investors holding cryptocurrencies that may not increase in value or, worse, may decrease in price.

So, where do you start your research? Consider this chapter as your crash course on cryptocurrency compliance. I employ the same methods as institutional investors when conducting compliance research for my clients. The regulatory analysis in these pages will put you in the same league as institutional investors, helping you make better and more informed cryptocurrency investment decisions.

GENSLER'S IMPACT ON CRYPTOCURRENCIES

Chairperson Gensler was a polarizing figure in the cryptocurrency community. Some cryptocurrency experts believe Chairperson Gensler over-regulated the cryptocurrency space, while others believe he is doing the right thing by protecting individual investors from fraud and scams. Regardless of one's opinion of Chairperson Gensler, during his tenure as SEC Chairperson, he exerted significant influence over the present and future of cryptocurrencies during a critical and formative period.

ATKIN'S IMPACT ON CRYPTOCURRENCIES

Chairperson Atkins has a very different approach to regulating cryptocurrencies than his predecessor, Chairperson Gensler. Atkins has publicly stated that his top goal as SEC Chairman is to create a framework, with the intention of reducing the ambiguity and lack of existing regulation that he believes creates uncertainty in the market and inhibits innovation. He has also expressed a desire to remove politics from the SEC's decision-making process regarding financial markets, including the regulation of cryptocurrencies.

During his Senate confirmation hearing, Atkins highlighted that providing a firm regulatory foundation for digital assets would be a "top priority." He believes that while creating regulations is essential, their successful implementation, with clear rules that benefit all market participants, is equally important.

The chapter quote from Chairperson Atkins of the Securities and Exchange Commission (SEC) strongly indicates that the SEC will continue to conduct investigations and bring charges against fraudulent cryptocurrencies and tokens that the SEC considers securities.

Overall, Atkins' perspective suggests a move towards a more strategic and less enforcement-heavy approach to crypto regulation compared to his predecessor. He is seen as pro-innovation and receptive to industry input, potentially leading to a more collaborative engagement between the SEC and the cryptocurrency sector.

IMPORTANT CRYPTOCURRENCY LEGAL ISSUES YOU SHOULD CONSIDER BEFORE YOU INVEST

Even if you are not in the legal profession, it is essential to understand the basic legal issues related to blockchain and cryptocurrency. Legal issues are present in all business transactions, including cryptocurrencies. Understanding these issues can help prevent you from making a poor investment decision and losing your investment dollars.

Before we dive into the intricacies of cryptocurrency regulations, here is a brief overview of what you will learn in this section.

1. **Regulatory Uncertainty:** Investors, particularly institutional investors, tend to prefer investing in industries with established regulatory frameworks. In contrast, cryptocurrencies are still operating in a state of regulatory uncertainty.
2. **Investor Protection:** Government agencies report increased cryptocurrency fraud, scams, and market manipulation, leading to financial losses for individuals.[57]
3. **Taxation Issues:** The taxation of cryptocurrencies is complex, and taxing authorities, including the IRS, are still determining tax policies for cryptocurrencies.

---

[57] Carmona, Tonantzin. Protecting the American Public from Crypto Risks and Harms. Brookings, 2025. https://www.brookings.edu/articles/protecting-the-american-public-from-crypto-risks-and-harms/#:~:text=According%20to%20the%20FBI%20Internet%20Crime%20Complaint,in%20losses%20compared%20to%20the%20previous%20year).&text=The%20FBI's%20IC3%20team%20reported%20nearly%2011%2C000,2024—a%2099%25%20increase%20from%20the%20previous%20year. (Accessed September 2, 2025)

4. **Know Your Customer (KYC) and Anti-Money Laundering (AML) compliance:** Cryptocurrency companies must comply with KYC and AML regulations by requiring users to provide proper identification before opening a wallet or account, and tracking countries that download the company's cryptocurrency wallets. Cryptocurrency companies are increasingly subject to AML and KYC regulations to prevent money laundering, terrorist financing, and other illicit activities. Compliance with these regulations will require cryptocurrency companies to invest in compliance software and experienced compliance resources.[58]
5. **Criminal Acts**: Investment fraud, including cryptocurrency investment fraud, is a serious crime. The Department of Justice is prosecuting individuals engaged in cryptocurrency investment fraud and Ponzi schemes.

The legal and regulatory landscape surrounding cryptocurrencies is rapidly evolving, and specific legal issues vary depending on the jurisdiction. Ensure you consult with legal professionals specializing in cryptocurrency law in the applicable countries to ensure compliance and mitigate legal risks.

CRYPTOCURRENCY REGULATION

I have discussed the importance of cryptocurrency regulations many times in the preceding chapters. Now, in this part, let's take a closer look at why regulation matters so much for investors and the future of cryptocurrency.

The regulation of cryptocurrencies presents numerous challenges due to their decentralized nature, technological advancements, and cross-border implications. All of these issues challenge regulators to keep up with the latest

[58] Howell, James. Cryptocurrency Compliance Explained. 101 Blockchains, July 25, 2022. https://101blockchains.com/cryptocurrency-compliance/ (Accessed September 2, 2025)

innovations, protect their constituents, and address potential risks. Some of the key complications and issues include:

1. **Classification**: One of the fundamental debates in cryptocurrency regulation is how to classify digital assets. Should they be treated as securities, commodities, currency, or something else entirely? This classification has significant implications for the cryptocurrency regulatory framework.
2. **Cross-Border Coordination:** Cryptocurrencies are essentially borderless. They operate on a global scale, making it difficult for individual countries to regulate them effectively. Coordinating regulatory efforts across different jurisdictions is essential to preventing the exploitation of regulatory system gaps and ensuring a level playing field. However, cross-border coordination has been rare so far, despite the need to ensure a level playing field.
3. **Consumer Protection:** Regulators are concerned about protecting consumers from fraud, scams, and market manipulation. Ensuring that investors have adequate information and safeguards is crucial but also challenging. In addition, new scams surface regularly, which adds to the challenges of consumer protection.
4. **Privacy and Surveillance**: Balancing the need for privacy with the requirement for transparency, Know Your Customer (KYC), and Anti-Money Laundering (AML) compliance is a delicate issue. Regulators must find ways to monitor cryptocurrency transactions without compromising individual privacy.
5. **Taxation:** Determining the appropriate tax treatment for cryptocurrency transactions is complex, as it involves issues of capital gains, income tax, and value-added tax (VAT).
6. **Market Manipulation:** Regulators face a significant challenge in preventing market manipulation involving cryptocurrencies.

I will be honest — this is a lot! However, these are all relevant issues that we must analyze to understand the risks associated with regulating and investing in cryptocurrencies. In addition, these are all important issues that all cryptocurrency investors should be aware of. As the industry continues to evolve, regulators will need to adapt their approaches and develop innovative solutions to address these challenges effectively.

The reality is that it takes a serious commitment of time and resources to not only learn about cryptocurrencies but also to stay updated on recent developments. Most politicians, regulators, and their staff have so much responsibility that they may not have time for a truly deep knowledge dive. Therefore, several legislative and policy decisions are arguably made without the benefit of advanced and in-depth knowledge.

President Biden and Trump both signed executive orders, including the establishment of task forces to conduct research and recommend actions to foster greater understanding of the use of cryptocurrencies. The next section will cover these executive orders.

THE WHITE HOUSE EXECUTIVE ORDER 14067

On March 9, 2022, President Biden signed Executive Order 14067, "Ensuring the Responsible Development of Digital Assets" (Digital Assets Executive Order). This Order coordinates the efforts of US government agencies to regulate and enforce the use of digital assets. [59]

The Digital Assets Executive Order was a step in the right direction; however, as with all executive orders, it had limitations. It is essential to note that executive orders are typically temporary. One Administration will

---

[59] Executive Office of the President. Ensuring Responsible Development of Digital Assets. Federal Register, March 9, 2022. https://www.federalregister.gov/documents/2022/03/14/2022-05471/ensuring-responsible-development-of-digital-assets (Accessed September 3, 2025)

implement the executive order, and four or eight years later, the next Administration may modify or even rescind it, potentially leaving a gap in the legal and regulatory process. These orders are often challenged in court, and in the process, the court may grant an injunction to stay the implementation of the executive order. To that point, President Trump rescinded Biden's Digital Assets Executive Order and implemented a replacement Executive Order.

THE WHITE HOUSE EXECUTIVE ORDER 14178

President Trump campaigned as a pro-cryptocurrency candidate, and so far, he has established a very progressive cryptocurrency agenda. President Trump rescinded President Biden's Executive Order 14067 and issued Executive Order 14178: Strengthening American Leadership in Digital Financial Technology (Digital Financial Technology Executive Order).[60]

A TALE OF TWO EXECUTIVE ORDERS

Executive Order 14067, issued by President Biden in 2022, focused on ensuring the responsible development of digital assets. It directed federal agencies to study the risks and benefits of digital assets, including cryptocurrencies, and to coordinate on policy recommendations.

Executive Order 14178, "Strengthening American Leadership in Digital Financial Technology" (also known as the 2025 Digital Asset Executive Order), takes a different approach. It is designed to strengthen American leadership in digital financial technology, emphasizing innovation and economic competitiveness. This Executive Order aims to streamline

---

[60] The White House. Strengthening American Leadership in Digital Financing Technology. Whitehouse.gov, January 23, 2025.https://www.whitehouse.gov/presidential-actions/2025/01/strengthening-american-leadership-in-digital-financial-technology/ (Accessed October 27, 2025)

regulation, protect individual freedoms, and foster the adoption of blockchain and cryptocurrency technologies.

Here's a side-by-side comparison of the two executive orders:

Feature	Executive Order 14067	Executive Order 14178
Focus	Responsible development of digital assets	American leadership in digital financial technology
Approach	Risk assessment and policy coordination	Streamlined regulation and innovation
Key themes	Risks and benefits, consumer protection, and financial stability	Innovation, economic competitiveness, and individual freedoms
Regulatory stance	Cautious and risk-averse	Pro-innovation and less regulatory

It's important to note that these are just two executive orders in a rapidly evolving cryptocurrency landscape. The specific policies and regulations related to digital assets are likely to continue to evolve in the years to come.

On March 9, 2025, President Trump signed Executive Order 14233, his second executive order related to cryptocurrency. Executive Order 14233 is entitled: Establishment of the Strategic Bitcoin Reserve and United States Digital Assets Stockpile (Bitcoin Reserve Executive Order).[61] Let's address this two-in-one Executive Order.

The Strategic Bitcoin Reserve is intended to be a permanent reserve asset composed of Bitcoin currently held by the U.S. government through criminal and civil asset forfeitures. The Department of the Treasury will

---

[61] The White House. Establishment of the Strategic Bitcoin Reserve and United States Digital Asset Stockpile. Whitehouse.gov, March 6, 2025. https://www.whitehouse.gov/presidential-actions/2025/03/establishment-of-the-strategic-bitcoin-reserve-and-united-states-digital-asset-stockpile/ (Accessed October 27, 2025)

manage Bitcoin strategically, with no plans to sell the Bitcoin in the reserve, and will explore budget-neutral strategies for future acquisitions. [62]

The United States Digital Asset Stockpile will consist of other digital assets, beyond Bitcoin, that have been forfeited to the Treasury. Unlike the Bitcoin Reserve, the Executive Order suggests that the Treasury may develop strategies for the responsible stewardship of this stockpile, including potential sales of the cryptocurrency. [63] Both initiatives require federal agencies to provide a comprehensive account of their digital asset holdings to the Treasury and the President's Working Group on Digital Asset Markets, marking a shift toward more centralized and strategic management of digital assets within the U.S. government.

Initially, the announcement of the Strategic Bitcoin Reserve did not impress the markets, with the price of Bitcoin even falling slightly. This suggests that the immediate impact on price might be muted until the reserve's operational details and future acquisition strategies become clearer. Over the long term, the decreased supply and potential rise in demand could lead to substantial price increases, although ongoing volatility in the cryptocurrency market is likely to persist.

I believe the Bitcoin Reserve Executive Order will have the following positive impact on Bitcoin and other cryptocurrencies.

- **Increased Legitimacy and Adoption**: The U.S. is a leader in the global cryptocurrency market, so the U.S.'s decision to create a Bitcoin Reserve and Digital Assets Stockpile could prompt other countries to create their own Bitcoin Reserve or consider similar strategies for managing digital assets, potentially leading to a domino effect of increased institutional involvement in Bitcoin.

---

[62] Ibid.

[63] Ibid.

- A U.S. Bitcoin reserve could also significantly enhance Bitcoin's perceived legitimacy as a digital asset and a store of value. This could encourage wider adoption by institutional investors, corporations, and even other nations.

- **Reduced Selling Pressure:** The Bitcoin Reserve Executive Order explicitly states that Bitcoin deposited into the Strategic Bitcoin Reserve will not be sold and will be maintained as a reserve asset. This removes a potentially significant source of selling pressure from the market, as the government's seized Bitcoin will now be held in a long-term investment.

- **Potential for Future Price Appreciation**: As Bitcoin and many other cryptocurrencies have a finite or limited supply, the mechanics of supply and demand are critical components of their price. If the U.S. government begins to strategically acquire more Bitcoin, this increased demand could positively influence Bitcoin's price over time, especially given its limited supply. Over the long term, the reduced supply and potential for increased demand could lead to significant price appreciation.

President Trump will be in office until 2028, and I believe he will continue to sign more cryptocurrency-based executive orders into law. That's why it's essential to stay up to date on any cryptocurrency-related executive orders.

FEDERAL LEGISLATION

When discussing cryptocurrency and blockchain federal legislation, a crucial point is the existence of the Congressional Blockchain Caucus, a bipartisan group of Members of Congress and Staff that aims to study blockchain technology and promote collaboration between industry and the government. When the Congressional Blockchain Caucus was founded in

September 2016, it held meetings with influencers in the cryptocurrency and blockchain space. I feel truly fortunate to have attended some of those early meetings, where I was genuinely impressed by the members' deep understanding of cryptocurrencies and blockchain technology. Their unwavering dedication to ensuring this innovative technology had the opportunity to grow and flourish—without the fear of being stifled by overbearing legislation—was truly inspiring. In a nation often divided along political lines, it was heartening to see members of both parties come together with openness and purpose to engage in meaningful, productive discussions about this exciting technology.

The US Congress has introduced numerous bills aimed at providing more clarity in the blockchain technology and cryptocurrency sector. Still, recently, only one bill passed in the US House and Senate to become law: The Guiding and Establishing National Innovation for U.S. Stablecoins Act of 2025 (GENIUS Act). [64]

Here's a summary of the GENIUS Act and its potential effects on individuals:

SUMMARY OF THE GENIUS ACT OF 2025:

The core purpose of the GENIUS Act is to establish a clear regulatory regime for stablecoins used for payment or settlement. The GENIUS Act defines a "payment stablecoin" as a digital asset designed to be used as a means of payment, where the issuer is obligated (or represents they will) to convert it for a fixed amount of monetary value. A key provision is the requirement for stablecoin issuers to hold 1:1 reserves of highly liquid assets (like U.S. dollars, short-term Treasury bills, or insured bank deposits) to back every stablecoin

[64] The Guiding and Establishing National Innovation for U.S. Stablecoins Act of 2025, S. 1234, 119th Cong., 1st sess. (2025).

issued. This is intended to ensure that stablecoins maintain their pegged value and can be redeemed at any time.[65]

- **Clarification on Securities Status**: The GENIUS Act explicitly states that payment stablecoins are not securities.
- **Consumer Protection:** The bill emphasizes consumer protection by:
  - Prohibiting misleading marketing: Issuers would be forbidden from misrepresenting stablecoins as being backed by the full faith and credit of the U.S. government or covered by FDIC insurance.
  - Transparency: Issuers would be required to provide monthly public disclosures of their reserve composition and, for larger issuers, annual audited financial statements.
- **Regulatory Oversight:** The Act creates a dual federal and state regulatory system:
  - Federal Oversight: Larger stablecoin issuers (typically those with over $10 billion in market capitalization) would primarily fall under the oversight of the Federal Reserve or the Office of the Comptroller of the Currency (OCC).
  - State Oversight: Smaller issuers (under $10 billion) could opt for state-level regulation, provided the state's regulatory regime is "substantially similar" to the federal framework.
- **Anti-Money Laundering (AML) and Sanctions Compliance:** Issuers would be treated as financial institutions for Bank Secrecy Act (BSA) purposes, requiring them to implement robust AML and sanctions compliance programs, monitor and report suspicious activity, and conduct enhanced due diligence.

---

[65] Stablecoin Legislation: An Overview of S.919, GENIUS Act of 2025. Congress.gov, May 19, 2025. https://www.congress.gov/crs-product/IN12522 (Accessed October 27, 2025)

- **Prohibition on Yield-Bearing Stablecoins:** The bill prohibits permitted stablecoin issuers from offering interest or yield on their stablecoins.[66]

HOW THE GENIUS ACT WILL AFFECT INDIVIDUALS:

In my view, the GENIUS Act could positively impact individuals in several ways:

- **Increased Confidence and Safety for Stablecoin Users:**
  - Reduced Risk of "De-pegging": The strict 1:1 reserve requirements and transparency measures are designed to make stablecoins more reliable and reduce the risk of them losing their value. Thus, individuals using stablecoins for payments or storing value can have confidence that their stablecoins are truly backed.
  - Better Protection in Case of Issuer Failure: The provision prioritizing stablecoin holders in bankruptcy provides increased consumer protection. If a stablecoin issuer goes bankrupt, individuals holding their stablecoins have a stronger claim to the underlying reserves.
  - Clearer Information: Monthly disclosures and audited financial statements provide individuals with more transparent information about the backing of their stablecoins.

- **Potential for Wider Adoption of Stablecoins:** With clear regulations and stronger consumer protections, more businesses and individuals may feel comfortable using stablecoins for remittances and everyday digital transactions.

In summary, the following story illustrates how the GENIUS Act can impact individuals.

---

[66] Ibid.

Sarah and Mark do not like cryptocurrencies; they find them too risky, and there are too many to choose from. One morning, Sarah noticed her favorite coffee shop started accepting stablecoins for payment. It was just as easy as swiping her card, but it seemed faster. Her friend, Mark, received his freelance payment instantly in the same stablecoin, avoiding bank transfer delays. While they didn't fully grasp the underlying mechanics, a newfound stability and ease in these digital transactions trickled into their everyday lives, making small purchases and peer-to-peer payments feel more seamless and secure than ever before. *Are you the next Sarah or Mark?*

FEDERAL REGULATORY ENFORCEMENT ACTIONS

At the time of publication, the cryptocurrency market is valued at $3.7 trillion, and all reputable sources predict it will continue to increase for years and decades. The fact that the cryptocurrency market is valued at trillions of dollars is extremely important to cryptocurrency regulation for several critical reasons, primarily related to financial stability, consumer protection, and the integration of crypto into the broader economy.

Here's a breakdown of why this massive market cap is so significant:

- **Systemic Risk to Financial Stability:**
    - "Too Big to Ignore": When an asset class reaches a multi-trillion-dollar valuation, it can no longer be dismissed as a niche market. It is large enough to impact traditional financial systems.
    - Interconnectedness: As the crypto market grows, its connections with traditional finance deepen. This includes:
        - Institutional Investment: More large financial institutions (banks, hedge funds, asset managers) are investing in crypto or offering crypto-related services. Bitcoin and Ethereum exchange-traded funds (ETFs) are a good example of interconnectedness.

  - Leverage and Lending: The use of leverage and lending within the crypto ecosystem can amplify shocks. If a major crypto entity collapses, its failure could ripple through the traditional financial system if there are significant interconnections.

- Contagion Risk: The collapse of large crypto entities (like FTX in 2022, or the de-pegging of TerraUSD) demonstrated how failures in one part of the crypto ecosystem can trigger widespread panic and losses, even if the direct impact on traditional banks was limited at that time due to existing "firewalls." As the market grows and connections deepen, such events could have a broader, more severe impact.

- **Consumer and Investor Protection:**

  - Lack of Safeguards: Unlike traditional financial markets, which have decades of established rules for investor protection (e.g., disclosure requirements, anti-fraud laws, deposit insurance, segregated customer funds), the crypto market has historically operated with far fewer safeguards. The large market capitalization highlights the urgency of implementing these protections.
  - Misleading Practices: The sheer size of the market means more opportunities for scams, deceptive marketing, pump-and-dump schemes, and entities falsely claiming government backing or insurance, as seen in past crypto failures.

- **Market Integrity and Transparency:**

  - Preventing Illicit Activity: A large, unregulated market is a greater magnet for illicit activities like money laundering, terrorist financing, and sanctions evasion. A $3.7 trillion market offers immense liquidity and scale for such activities, making it a priority for regulators to implement robust Anti-Money Laundering (AML) and Know Your Customer (KYC) requirements.

- Price Manipulation: Without oversight, large market participants could more easily engage in price manipulation, harming smaller investors. The substantial market size makes this a more significant concern.

The $3.7 trillion (and growing) valuation of the cryptocurrency market transforms it from a fringe curiosity into a major financial force requiring smart and balanced laws and regulations to ensure stability, consumer protection, and maintain the integrity of the broader financial system.

The next section of the chapter will cover U.S. regulatory enforcement in the following order: the Securities and Exchange Commission (SEC), the Commodity Futures Trading Commission (CFTC), the Department of Justice (DOJ), the Department of the Treasury (Treasury), and the Internal Revenue Service (IRS).

JOINT STATEMENT ON ACTIVITIES INVOLVING DIGITAL ASSETS UNDER THE BIDEN ADMINISTRATION

In 2019, the SEC was the most active federal agency in terms of cryptocurrency enforcement actions and litigation. Since 2019, the Commodity Futures Trading Commission, the Department of the Treasury, and the Department of Justice have significantly increased their enforcement actions against cryptocurrency-related offenses. In October 2019, the SEC, CFTC, and the Treasury FinCEN division issued the Joint Statement on Activities Involving Digital Assets (Joint Statement on Digital Assets). This Statement communicated the importance of government agencies working together to regulate and investigate cryptocurrency companies, specifically in relation to anti-money laundering and countering the financing of terrorism (AML/CFT) obligations under the Bank Secrecy Act (BSA).

The Joint Statement on Digital Assets provides important information outlining the regulatory requirements depending on the type of digital asset:

> The nature of the digital asset-related activities a person engages in is a key factor in determining whether and how that person must register with the CFTC, FinCEN, or the SEC. For example, certain "commodity"-related activities may trigger registration and other obligations under the Commodity Exchange Act (CEA), while certain activities involving a "security" may trigger registration and other obligations under the federal securities laws. If a person falls under the definition of a "financial institution," its AML/CFT activities will be overseen for BSA purposes by one or more of the Agencies (and potentially others). For example, the AML/CFT activities of a futures commission merchant will be overseen by the CFTC, FinCEN, and the National Futures Association (NFA); those of an MSB (money service business) will be overseen by FinCEN; and those of a broker-dealer in securities will be overseen by the SEC, FinCEN and a self-regulatory organization, primarily the Financial Industry Regulatory Authority (FINRA).[67]

The Joint Statement on Digital Assets guidance is important for cryptocurrency companies and investors. Cryptocurrency companies issuing a cryptocurrency or token are advised to use this joint statement as a guide for their regulatory efforts and consult with an experienced cryptocurrency attorney.

As an investor, you can use this information as a guide to determine where a cryptocurrency should be registered before deciding to invest in it.

- If the cryptocurrency is a security, it should be registered with the SEC.

---

[67] Tarbert, Heath; Blanco, Kenneth; and Clayton, Jay. "Leaders of CFTC, FinCEN, and SEC Issue Joint Statement on Activities Involving Digital Assets." October 11, 2019. https://www.sec.gov/news/public-statement/cftc-fincen-secjointstatementdigitalassets. (Accessed September 2, 2025)

- If the cryptocurrency is a commodity, it should be registered with the CFTC and NFA.
- If the cryptocurrency is used as money, it should be registered with FinCEN.
- Most importantly, the cryptocurrency may fall under all three categories; if that's the case, it should be registered with all of the above regulatory bodies.[68]

TRUMP ADMINISTRATION AND CRYPTOCURRENCIES

Under the Biden Administration, the SEC, CFTC, and Treasury FinCEN divisions were working together to comprehensively regulate cryptocurrencies and enforce actions in the cryptocurrency space. The Trump Administration is taking a different approach.

JOINT STATEMENT ON ACTIVITIES INVOLVING DIGITAL ASSETS UNDER THE TRUMP ADMINISTRATION

On September 2, 2025, the CFTC and SEC issued the following statement: CFTC and SEC Staff Issue Joint Statement on Trading of Certain Spot Crypto Asset Products (Joint Statement on Spot Crypto Assets). This joint statement dramatically changes the CFTC and SEC crypto regulatory approach from the Biden Administration, with the intent to make the US the "crypto capital of the world".[69]

In the Joint Statement, Chairperson Atkins states:

---

[68] Statement. Leaders of CFTC, FinCEN, and SEC Issue Joint Statement on Activities Involving Digital Assets. SEC.gov, October 11, 2019. https://www.sec.gov/news/public-statement/cftc-fincen-secjointstatementdigitalassets (Accessed October 27, 2025)

[69] Press Release. CFTC and SEC Staff Issue Joint Statement on Trading of Certain Spot Crypto Asset Products. CFTC.gov, September 2, 2025. https://www.cftc.gov/PressRoom/PressReleases/9112-25 (Accessed September 4, 2025)

"Today's joint staff statement represents a significant step forward in bringing innovation in the crypto asset markets back to America. Market participants should have the freedom to choose where they trade spot crypto assets. The SEC is committed to working with the CFTC to ensure that our regulatory frameworks support innovation and competition in these rapidly evolving markets."[70]

To be clear, the Joint Statement on Spot Crypto Assets is a crucial piece of guidance that suggests a more welcoming and collaborative regulatory environment. It reduces legal uncertainty, potentially opens the door for traditional exchanges to enter the crypto market, and signals a move toward a more pro-crypto and integrated digital asset ecosystem in the U.S.

The next step in this regulatory analysis is to review the Trump Administration's choices to lead the following agencies: the Securities and Exchange Commission, the Commodities and Futures Trading Commission, and the Department of the Treasury.

PAUL ATKINS, CHAIRPERSON, SECURITIES AND EXCHANGE COMMISSION

Paul Atkins is the Trump-appointed SEC Chairperson. He was the Co-Chairperson of the Token Alliance and a member of the Board of Advisors to the Chamber of Digital Commerce, bringing private-sector experience to his new role as SEC Chairperson.

Unlike former Chairperson Gensler, who many viewed as too aggressive against cryptocurrency companies, Chairperson Atkin's previous crypto

[70] Ibid.

experience, pro-crypto stance, and statements have been welcomed by the crypto community. [71]

SCOTT BESSENT, SECRETARY, THE DEPARTMENT OF THE TREASURY

Scott Bessent, the Treasury Secretary, is a billionaire hedge fund manager and a crypto advocate who has been welcomed by the crypto community. As Treasury Secretary, Bessent will serve on the Presidential Working Group on Digital Asset Markets to design a U.S. government strategy for digital assets. Unlike his predecessor, Janet Yellen, who was open to a US CBDC, Secretary Bessent made it clear very early on that he does not support a US CBDC when he stated, "In my mind, a central bank digital currency is for countries that have no other investment alternatives."[72]

As the IRS falls under Treasury as well, Secretary Bessent has authority over the taxation of cryptocurrencies. The cryptocurrency tax policy decisions Bessent makes, or does not make, will have a significant impact on institutional and individual cryptocurrency investors.

BRIAN QUINTENZ, THE CHAIRPERSON OF THE COMMODITIES AND FUTURES TRADING COMMISSION

Before becoming the Chairperson of the Commodity Futures Trading Commission, he was the Head of Policy for a16z Crypto. However, this is not Commissioner Quintenz's first time at the CFTC, as he has served under both Presidents Obama and Trump as a Commissioner of the Commodity Futures Trading Commission (CFTC). During his service, the CFTC oversaw the listing of the first U.S.-regulated Bitcoin and Ether futures

---

[71] Dobkin, Rachel. Trump Taps Crypto Advocate Paul Atkins for SEC Chair. Newsweek, December 4, 2024. https://www.newsweek.com/donald-trump-paul-atkins-sec-cryptocurrency-1995663 (Accessed October 26, 2025)

[72] Hamilton, Jessee. Trump Treasury Pick Bessent Opposes Idea of a US Central Bank Digital Currency. Coindesk.com, January 16, 2025. https://www.coindesk.com/policy/2025/01/16/trump-treasury-pick-bessent-opposes-idea-of-u-s-central-bank-digital-currency (Accessed October 27, 2025).

contracts on derivatives exchanges. Brian Quintenz has made it clear that under his leadership, the CFTC will support cryptocurrencies and blockchain technology, as he stated on 𝕏: "The agency (CFTC) is also well poised to ensure the USA leads the world in blockchain technology and innovation."[73]

SECURITIES AND EXCHANGE COMMISSION (SEC) REGULATION OF CRYPTOCURRENCIES

The cryptocurrency industry is often compared to the Wild West: a rapidly evolving frontier with immense potential and just as many risks and dangers. As Chairperson Atkins has replaced Chairperson Gensler, it can be said literally, "there is a new sheriff in town." As the digital equivalent of local law enforcement, the SEC has taken on the challenging task of regulating this new industry and protecting investors.

The SEC's primary goal is to protect investors from fraud and manipulation while fostering an environment conducive to innovation. [74]However, the blurred lines between traditional securities and digital assets have presented many challenges.

To navigate this complex terrain, the SEC relies on existing securities laws, specifically the Securities Act of 1933 [75]and the Securities Exchange Act of 1934[76]. This approach has led to a series of high-profile enforcement

[73] Brian Quintenz. 𝕏 @Brian Quintenz. February 12, 2025. https://x.com/BrianQuintenz/status/1889702678852211005 (Accessed October 29, 2025)

[74] About. SEC.gov. https://www.sec.gov/about#:~:text=The%20SEC's%20mission%20is%20to,markets;%20and%20facilitate%20capital%20formation. (Accessed September 30, 2025)

[75] Securities Act of 1933, 15 U.S.C. §§ 77a et seq. (1933).

[76] Securities Exchange Act of 1934, Public Law 291, 73rd Cong., 48 Stat. 881 (1934), codified as amended at 15 U.S.C. §§ 78a–78nn.

actions and regulatory guidance, shaping the crypto landscape. Most of these enforcement actions start with the Howey Test analysis.

THE HOWEY TEST

The "Howey Test" is one of the tools the SEC uses to determine if a transaction qualifies as an investment contract that should be regulated as a security[77]. It is based on the Supreme Court case SEC v. Howey Co., in which the Howey Co. sold parts of its citrus land to raise capital to finance additional land development. The SEC filed an injunction against Howey Co. because it did not register the land transactions. The Supreme Court found Howey Co. violated securities laws because the purchasers of the land were spectators (passive investors) who did not possess the "knowledge, skills, and equipment necessary for the cultivation of citrus trees." As such, the land purchasers were investing in Howey Co.'s expertise to improve the citrus land, thereby increasing the value of their investment.[78] The Howey Test can apply to any industry and to any investment.

The SEC uses three steps to apply the Howey test:

1. **The Investment of Money**—The digital asset is acquired in exchange for value.
2. **Common Enterprise**—A business or investment that is managed by third parties.
3. **Reasonable Expectation of Profits Derived from the Efforts of Others**—This is exactly what it sounds like: participants have an

---

[77] SEC v. Howey Co., 328 U.S. 293 (1946).

[78] Ibid.

expectation of return on investment based on someone else's effort or expertise.[79]

How does the Howey Test apply to cryptocurrencies? Overall, the Howey Test is relevant to all investments. Suppose a cryptocurrency investment is speculative and the investor relies on the management team's expertise to enhance its value. In that case, the SEC is likely to classify the cryptocurrency as an investment subject to SEC regulations.

When I use the Howey Test to determine if a potential client's cryptocurrency or token is a security and recommend that they register their cryptocurrency with the SEC, I am often asked the same question: What is the benefit of filing our cryptocurrency, token, or ICO with the SEC? I have been asked this question so many times that I have designed a standardized response.

- First, investors will want to know that you registered your cryptocurrency or token as a security with the SEC. An SEC filing reassures investors that you are in compliance with the securities laws.
- Second, you do not want to be on the wrong side of an SEC enforcement action. If the SEC rules against you, you may be penalized and required to repay the investors.
- Finally, why do you file taxes every year? The benefit is compliance with tax laws and avoiding trouble with the IRS. The same principle applies to registering your cryptocurrency or token with the SEC.

In 2024, I gave a speech at the Atlanta Blockchain Center about cryptocurrency legal issues. I asked the audience the following question: If you were presented with two very similar cryptocurrencies to consider for

---

[79] Securities and Exchange Commission. "Framework for 'Investment Contract' Analysis of Digital Assets." SEC.gov, April 3, 2019, https://www.sec.gov/corpfin/framework-investment-contract-analysis-digital-assets. (Accessed September 30, 2025).

investment, one registered with the SEC and the other not registered, which cryptocurrency would you choose to invest in? Every audience member responded that they would invest in the cryptocurrency registered with the SEC. The credibility of regulatory registration is a crucial factor that every cryptocurrency company and investor should strongly consider.

The SEC v. Ripple Labs case, covered in the next section, is a significant case for applying the Howey test to cryptocurrencies.

SECURITIES AND EXCHANGE COMMISSION V. RIPPLE LABS

Ripple Labs, Inc., founded in 2012, uses its Ripple Network to provide financial institutions with low-cost, expedited cross-border payment clearance. In December 2020, the SEC sued Ripple Labs in the U.S. District Court for the Southern District of New York. In Securities and Exchange Commission v. Ripple Labs, Inc., Bradley Garlinghouse, and Christian A. Lawsen (SEC v. Ripple), the SEC alleged that Ripple Labs had an initial public offering (IPO), also known as an initial coin offering (ICO). Ripple's IPO raised about $600 million by selling XRP tokens. The SEC argued that Ripple did not register the XRP tokens as a security or file for an exemption from registration. Ripple's defense was that XRP was not a security, so it was not required to file XRP with the SEC or seek an exemption. Usually, when the SEC files charges against or fines a company for selling an unregistered security, the company pays the fine or settles with the SEC. Since Ripple did not pay the fine or settle, the SEC sued Ripple Labs.[80]

The main legal issue in the Ripple case was the common enterprise element of the Howey test. The SEC argued that XRP was a security because it was an investment of money in a common enterprise, with an expectation

---

[80] SEC Charges Ripple and Two Executives with Conducting $1.3 Billion Unregistered Securities Offering. December 22, 2020. https://www.sec.gov/newsroom/press-releases/2020-338 (Accessed October 28, 2025)

of profits derived from the efforts of Ripple Labs. Ripple Labs argued that XRP was a cryptocurrency, not a security, and was not subject to securities regulation. According to the XRP functioned as a digital asset used for efficient cross-border transactions.[81]

The SDNY court issued a two-fold ruling. One part involved XRP tokens purchased by individual investors, while the other involved contractual sales of XRP tokens to institutional investors. Regarding the XRP tokens to individual investors, the SDNY Court ruled that Ripple was not liable under federal securities law for selling XRP tokens on the open market because the individuals who purchased XRP on a cryptocurrency exchange did not have an "investment contract" with Ripple. The Court reasoned that individual investors did not know that Ripple was on the other side of the trades, so they could not expect Ripple to use XRP sales revenue to boost the value of XRP tokens.[82]

Regarding institutional investors, the SDNY court denied Ripple's request for summary judgment and found that Ripple had investment contracts with these investors; therefore, the SDNY Court may still consider those XRP tokens as securities.

The Ripple motion to dismiss decision had a significant impact on the cryptocurrency market in several ways.

1. The Ripple summary judgment makes it more difficult for people who purchase cryptocurrencies from an exchange to argue that such purchases, without an investment contract, are unregistered securities.

---

[81] Nessi, Lorena, Dr. Ripple vs. SEC Lawsuit: A Complete Timeline of the Crypto Legal Battle. CCN.com, August 5, 2025. https://www.ccn.com/education/crypto/ripple-vs-sec-timeline-and-outcomes/ (Accessed October 17, 2025)

[82] SEC v. Ripple Labs, Inc., 682 F. Supp. 3d 308 (S.D.N.Y. 2023)

2. Companies that sell cryptocurrencies or tokens without an investment contract have a defensible position not to register their cryptocurrency or token with the SEC.

In August 2024, the SDNY court released the final judgment in SEC v. Ripple, ruling that Ripple's XRP institutional sales constituted securities transactions. Ripple was ordered to pay a $125 million civil penalty and is prohibited from selling unregistered XRP. In effect, Ripple must register XRP as a security with the SEC if it wishes to continue selling XRP.[83]

The SEC v. Ripple decisions established a federal court precedent for determining whether cryptocurrencies are securities or commodities. Many of us in the legal profession were waiting for both the SEC and Ripple to appeal the SDNY decisions. However, when Trump became President, it became clear that the SEC would not appeal the SDNY decision. As expected, in March 2025, the SEC moved to end its long-running lawsuit against Ripple Labs, withdrawing its appeal and bringing the four-year case to a close while leaving the court's prior rulings, including the $125 million penalty and injunction, in place.

The SEC's more accommodating approach to crypto enforcement has been welcomed by much of the cryptocurrency community (Reuters). As the agency pivots toward clearer rulemaking, the question now is: what does this shift mean for ICOs and other token offerings?

ARE ICOS DEAD?

The Ripple case is an example of the SEC regulating and, at times, filing lawsuits against companies that offer ICOs. The answer to this question depends on your perspective. For myself and other like-minded investors,

---

[83] *Final Judgment Order*, SEC v. Ripple Labs, Inc., *No. 20-cv-10832 (S.D.N.Y. Aug. 7, 2024), ECF No. 973 (the "Order").*

ICOs were never "alive" or a viable investment option. I always felt that the risks of investing in an ICO outweighed the benefits, and I was justified in that feeling, as none of the ICOs with which I had first-hand knowledge were successful in the long term. Eventually, all of them failed. However, as companies continue to create ICOs, they are not "dead." Following the failure of most ICOs and the SEC's significant increase in ICO enforcement, investors became more cautious with their investment dollars.

Case in point, in August 2025, ICO Bench published an article titled "10 Ethereum ICOs and Presales in August 2025".[84] This article, along with similar ones, demonstrates that ICOs are not dead. ICOs were a novel way to raise capital in the cryptocurrency space, but often lacked regulatory oversight and transparency, raising concerns about investor risk. Following the ICO craze, Security Token Offerings (STOs) emerged as a new method for cryptocurrency companies to raise capital. STOs are often described as the 'new ICOs', but with key differences:

Aspect	ICOs (Initial Coin Offerings)	STOs (Security Token Offerings)
Underlying Asset	Usually very speculative, representing a new cryptocurrency or project	Usually very speculative, representing a new cryptocurrency or project
Regulatory Oversight	Minimal (growing scrutiny)	Stronger compliance (e.g., SEC, KYC, AML)
Investor Protection	Limited, higher risk of fraud	Greater protection through regulation, legal recourse
Investor Access	Open to anyone with cryptocurrency	Often restricted to accredited investors

[84] Layram, Ruby. 10 Best Ethereum ICOs and Presales in September 2025. ICO Bench, August 28, 2025. https://icobench.com/cryptocurrency/ethereum-icos/?source=googleads&utm_source=google&utm_medium=cpc&gad_source=1&gad_campaignid=22610320291&gbraid=0AAAAA-tnDqZFZMaQuBCds59oQ5AKxp8RG&gclid=Cj0KCQjw_L_FBhDmARIsAItqgt55LL1ZAIHYryywV0232E1p4GJeVmJ50netwr4Y2a2Jj_7YDoLZGtoaAjroEALw_wcB (Accessed September 30, 2025)

Aspect	ICOs (Initial Coin Offerings)	STOs (Security Token Offerings)
Fundraising Speed	Faster, can be done quickly	Slower due to regulatory requirements
Token Type	Utility tokens (non-equity)	Security tokens (equity/ownership, profit-sharing)
Costs	Lower costs	Higher costs due to legal and compliance requirements
Transparency	Less transparent, more risk	More transparent, regulated environment
Liquidity	High volatility, no guaranteed liquidity	Potential liquidity on regulated exchanges
Investor Returns	Speculative, no equity or dividends	Equity, dividends, or profit-sharing rights
Risk	Higher risk due to speculative nature	Risk is still present but lower due to higher regulatory compliance and backing by real-world assets

In essence, STOs are designed to bring the benefits of blockchain technology to traditional securities markets, offering potential returns to investors while adhering to a more regulated framework.

THE TRUMP ADMINISTRATION AND CRYPTOCURRENCIES

Given President Trump's progressive cryptocurrency agenda, the SEC has adopted a different approach to regulating cryptocurrencies. Commissioner Atkins, the new Trump-appointed SEC Commissioner, showed his vision for the SEC when he stated, "If the SEC were more accommodating and would deal straightforwardly with these various firms, I think it would be a lot better" for crypto companies to work directly in the US instead of going overseas.[85] Chairperson Atkins is still relatively early in his tenure as SEC Chairperson. Still, it appears that the SEC's approach under Chairperson

85 Ross, Katherine. What SEC Chair Nominee Paul Atkins Has Previously Said About Crypto. Blockworks.co, December 5, 2024. https://blockworks.co/news/sec-chair-nominee-crypto-past (Accessed October 27, 2025)

Atkins is to collaborate with cryptocurrency companies, which may result in fewer investigations and lawsuits.

HISTORY OF DEREGULATION AND ITS IMPACT ON FINANCIAL MARKET CRISES

The Trump Administration, especially the SEC, CFTC, and Treasury, is already reducing investigations and enforcement actions, and possibly rolling back rules and regulations enacted during the Biden Administration. This regulatory rollback may encourage innovation and increase cryptocurrency prices; however, it also presents significant risks.

History has a way of repeating itself, and there is a history of the effect of financial deregulation on market prices. In the 1970s, a wave of deregulation swept through many economies, driven by the belief that less government intervention would lead to greater efficiency and economic growth. But there is an argument that financial deregulation led to the following market crashes:

Savings and Loan Crisis (1980s): Deregulation of the savings and loan industry led to risky lending practices, ultimately resulting in widespread failures and a massive taxpayer bailout.[86]

Financial Crisis of 2008: The repeal of the Glass-Steagall Act, which separated commercial and investment banking, allowed for increased risk-taking and interconnectedness within the financial system. The

[86] Kenton, Will. Understanding the Savings and Loan Crisis: Key Events and Its Impact. Investopia.com. August 26, 2025. https://www.investopedia.com/terms/s/sl crisis.asp#:~:text=The%20roots%20of%20the%20S&L%20crisis%20lay,knew%20of—and%20allowed—such%20fraudulent%20transactions%20to%20happen. (Accessed October 27, 2025)

deregulation of the mortgage market fueled the subprime mortgage crisis, leading to widespread defaults and a systemic collapse.[87]

The Savings and Loan Crisis and the Financial Crisis of 2008 both show that reduced oversight can encourage financial institutions to engage in more speculative and riskier activities, prioritizing short-term profits over long-term stability. When institutions believe they are "too big to fail," they may take on excessive risks, knowing that the government will likely intervene to prevent their collapse. Also, consider that the 'too big to fail' justification for a government to bail out a company may not apply to cryptocurrency companies, as none of them are currently considered too big to fail.

Moreover, how can the government save a decentralized company? For example, how can a government rescue Bitcoin when no one knows who actually created it, and when it is not run by any centralized organization? So, which government would rescue Bitcoin, given that it is a global cryptocurrency rather than a fiat currency?

In addition to riskier activities, deregulation can increase interconnectedness within the financial system, meaning that the failure of one institution can have a domino effect, triggering a broader crisis. Ultimately, deregulation can erode consumer protections, leaving individuals susceptible to predatory lending practices and financial exploitation.

---

[87] Maverick, J.B. Consequences of the Glass-Steagall Act Repeal. Investopia.com, October 05, 2023. https://www.investopedia.com/ask/answers/050515/did-repeal-glasssteagall-act-contribute-2008-financial-crisis.asp (Accessed October 27, 2025)

THE POTENTIAL IMPACT OF REDUCED OVERSIGHT ON THE CRYPTOCURRENCY MARKET

Because cryptocurrencies are based on technology and many of them are decentralized, deregulation would impact the cryptocurrency market differently than the Savings and Loan Crisis and the Financial Crisis of 2008.

Reduced regulatory oversight could foster innovation in the cryptocurrency space, leading to the development of new and potentially beneficial technologies. These new technologies may start small, but they have the potential to grow into powerful and profitable entities. Reduced barriers to entry could increase competition within the cryptocurrency market, potentially benefiting consumers.

Reduced regulatory oversight also presents significant risks, as it could lead to even greater volatility in the cryptocurrency market, particularly if speculation and risk-taking increase. This risk is compounded by the highly interconnected nature of the cryptocurrency market. The failure of one major exchange or stablecoin could trigger a cascading effect. For example, although it was not directly caused by deregulation, the 2021 cryptocurrency crash began with the de-pegging of TerraUSD (UST), a stablecoin.

Finally, the integration of cryptocurrencies into the traditional financial system could pose systemic risks if not properly regulated. One example of this integration is the cryptocurrency ETFs. According to VettaFi, as of the time of publication, 39 cryptocurrency ETFs are already integrated into the traditional financial system. Suppose deregulation, or even reduced regulatory enforcement, triggers a chain of events that causes the cryptocurrency market to crash. In that case, the cryptocurrency ETFs will likely crash as well, which would impact the traditional financial system.[88]

FINDING THE RIGHT BALANCE

The irony of this section is that Satoshi Nakamoto created Bitcoin in response to the 2008 Financial Crisis, and in 2025, the global cryptocurrency market cap reached an all-time high of $4.2 trillion USD. Has the solution to a financial crisis become the financial institution that will be at the center of the next financial crisis? If this is possible, which it is, how can a future cryptocurrency financial crisis be avoided?

The key to avoiding the next financial crisis lies in striking the right balance between regulation, oversight, and innovation. Overly strict regulations could stifle growth and innovation, while insufficient regulation could pose significant risks to the financial system. Regulations for cryptocurrencies should be tailored to the specific risks and characteristics of this emerging market. Finally, given the global nature of cryptocurrencies, international cooperation is crucial to ensure effective regulation and prevent regulatory arbitrage.

CRYPTOCURRENCIES, CELEBRITIES, AND THE SEC

The SEC has also increased enforcement against celebrities and influencers who endorse or promote cryptocurrency companies and investment offerings. Consequently, several celebrities and athletes have faced legal issues related to their cryptocurrency endorsements.

---

[88] Cryptocurrency ETF List. VettaFi. ETFDB.com, https://etfdb.com/etfs/currency/cryptocurrency/#etfs&sort_name=assets_under_management&sort_order=desc&page=2 (Accessed October 27, 2025)

KIMBERLY KARDASHIAN

Are you keeping up with the Kardashians? If you are, then you know that Kim Kardashian has had dealings with the SEC. In the Matter of Kimberly (Kim) Kardashian, the SEC charged Kim Kardashian for promoting EMAX tokens on social media without disclosing the $250,000 payment she received for the promotion. Kardashian agreed to settle the charges, pay $1.26 million in penalties, give up her EMAX tokens, pay any interest earned, and cooperate with the SEC's investigation.[89]

FLOYD MAYWEATHER JR. AND D.J. KHALED

In 2018, retired world champion boxer in multiple weight classes, Floyd Mayweather Jr., and musician DJ Khaled were separately fined by the SEC for promoting a cryptocurrency called Centra Tech without disclosing that they had been paid to do so. The SEC alleged that both men separately promoted Centra Tech's initial coin offering on social media, including 𝕏 and Instagram, without disclosing that they had been paid to do so. Mayweather agreed to pay over $600,000 in fines and make a repayment to the SEC to settle charges related to his promotion of Centra Tech, and to refrain from promoting any securities for a period of three years. On the same charges, DJ Khaled agreed to pay $150,000 in fines and repayments and agreed not to promote any securities for a period of 2 years.[90]

Celebrity and influencer endorsements can artificially inflate a cryptocurrency's price in the short term, leading to a sudden drop once the hype subsides. This is often part of a "pump-and-dump" scheme. For this and

---

[89] Press Release. SEC Charges Kim Kardashian for Unlawfully Touting Crypto Security. SEC.gov, October 3, 2022. https://www.sec.gov/newsroom/press-releases/2022-183#:~:text=The%20Securities%20and%20Exchange%20Commission,%22Ms. (Accessed October 28, 2025)

[90] Securities and Exchange Commission. Two Celebrities Charged With Unlawfully Touting Coin Offerings. SEC.gov, November 29, 2018. https://www.sec.gov/news/press-release/2018-268 (Accessed January 5, 2026).

many other reasons, investors should avoid making impulsive decisions driven by hype or the fear of missing out (FOMO).

Celebrities and influencers who accept money to endorse or promote questionable ventures can face legal and financial consequences. They must do their due diligence and disclose any potential conflicts of interest. This is because while celebrity endorsements and investments can be lucrative, they also come with significant risks.

Most celebrities and influencers lack proper expertise in financial markets (readers are advised to do their own research to confirm this). If you rely solely on celebrity endorsements to invest in a cryptocurrency, you may be putting your hard-earned money into a risky asset based on the statements or actions of someone who has no financial expertise. It is also important to remember that celebrities are typically well compensated for their endorsements, meaning their support is often not an objective opinion but simply a business arrangement. Moreover, celebrities and influencers are unlikely to be accountable for any losses incurred by investors. Regardless of whether the cryptocurrency decreases in price or even goes out of business, they are still paid for their endorsement. The only one to lose money will be the uninformed investor!

The next series of cases shows the importance of thorough research before investment.

THE $2 BILLION CRYPTO LENDING PROGRAM FRAUD

SCENARIO: You are interested in investing in cryptocurrencies, so you research cryptocurrency companies on social media. You watch YouTube, TikTok, or Instagram video testimonials from people who claim to have made a significant amount of money buying and selling a specific cryptocurrency. The videos have thousands of likes and hundreds of positive comments. You begin considering investing in this cryptocurrency to make extra money. What are your next steps?

BlockFi was a cryptocurrency lending and financial services company that allowed users to lend, borrow, and earn interest on digital assets.[91] In re BlockFi Lending LLC, the SEC charged BlockFi for failing to register the offer and sale of its crypto lending product (BlockFi Interest Accounts). And in a separate SDNY case, the SEC charged five individuals for allegedly promoting BitConnect's unregistered crypto offering, which it said raised about $2 billion worldwide. The SEC complaint states that these individuals marketed and sold unregistered securities in the Bitconnect lending program. The promoters advertised Bitconnect's lending program to investors by posting "testimonial" style videos on YouTube and earning commissions for successfully soliciting investor funds.

BlockFi, without admitting any wrongdoing, agreed to pay $100 million and register its cryptocurrency lending product with the SEC.[92] BlockFi ultimately filed for bankruptcy.

LESSONS LEARNED 1: Don't take everything you see or read about cryptocurrencies on social media at face value. Sometimes, testimonial videos can be misleading and might be designed to lure your investment dollars. Before you decide to invest, it's a good idea to do some research on government websites and rely on objective, reputable social media sources. If a cryptocurrency company's testimonial is pushing you to invest quickly, that's a red flag to watch out for.

---

[91] Marks, Alex. What Happened to BlockFi? Bankruptcy Explained in Simple Terms. Lendn. June 29, 2024. https://www.ledn.io/post/what-happened-to-blockfi (Accessed October 31, 2024)

[92] "BlockFi Agrees to Pay $100 Million in Penalties and Pursue Registration of its Crypto Lending Product." U.S. Securities and Exchange Commission, February 14, 2022, https://www.sec.gov/news/press-release/2022-26 (Accessed September 19, 2024)

LESSON LEARNED 2: BlockFi's failure to register its interest-bearing lending product with the SEC highlighted regulatory risks for investors, who did not receive the protections and disclosures required for registered securities.

THE $124 MILLION CRYPTOCURRENCY FRAUD OPERATION

SCENARIO: You are a cryptocurrency investor, and you learn about a token offering that raised $8 million. You research the token and discover that it has a sophisticated website and a strong social media presence. You also follow the token on cryptocurrency exchanges and notice that it has a good market value and trading volume. Now, should you invest in the token? The following SEC Enforcement Action illustrates how investing in this token can result in the loss of most, if not all, of your investment.

NASGO claimed to have developed a blockchain-based platform that enables clients to tokenize their companies, launch applications and databases, and utilize the NASGO network for global e-commerce transactions. It also sold digital asset securities called NSG tokens to the public as unregistered securities.[93]

NASGO allegedly raised $10 million through two fraudulent and unregistered securities offerings. SEC alleges NASGO "lied about the volume of NSG tokens sold, the number of users subscribed to NASGO's platform, and the project value of the tokens." When the interest in NSG tokens decreased, NASGO created Sharenode and used it to develop SNP tokens, another unregistered security. NASGO claimed that SNP tokens were on the NASGO blockchain. They also alleged the SNP tokens "would increase in

[93] Litigation Release No. 25377. SEC Alleges Fraud in Digital Asset Securities Offerings. SEC.gov, April 28, 2022. https://www.sec.gov/enforcement-litigation/litigation-releases/lr-25377 (Accessed September 19, 2025)

value by $.10 each week and $.10 for each new business that joined the NASGO platform."

According to the SEC, NASGO misappropriated nearly $4 million of investor funds, used 133 Bitcoin to list NSG on an unregistered trading platform, and funded a group of traders to trade NSG tokens amongst themselves. NASGO's fraudulent trading scheme gave a false impression that NSG tokens had a positive market share and price due to investor demand. The NASGO and NSG investment scam fell apart when investors outside the group of funded traders tried to sell their NSG tokens and found no buyers. When the NASGO traders stopped trading, the NSG token's value virtually disappeared, and investors lost their investment.[94]

LESSON LEARNED 1: It is a good investment practice to research whether a cryptocurrency or crypto-token is registered as a security with the SEC. Any cryptocurrency company dedicated to compliance will register its cryptocurrency with the SEC.

LESSON LEARNED 2: If it is too good to be true, it is probably a scam. Pay close attention to marketing and communications with investors. The statement that the SNP tokens "would increase in value by $.10 each week and $.10 for each new business that joined the NASGO platform" is a huge red flag. Any investment that guarantees a specific rate of return over a specified period should be avoided. Moreover, the investment rate of return may be closely related to a Ponzi scheme if, as in the NASGO case, the payout

---

[94] Litigation Releases. "SEC Alleges Fraud in Digital Asset Securities Offerings." U.S. Securities and Exchange Commission, April 28, 2022. https://www.sec.gov/litigation/litreleases/lr-25377#:~:text=Chiang%20and%20Tippetts%20then%20allegedly,projected%20value%20of%20NSG%20tokens. (Accessed September 19, 2025)

is based on '$.10 for each new business that joins the NASGO platform'. Another huge red flag!

VIDEO SHARING DIGITAL ASSET CREDITS

SCENARIO: You choose to invest in a cryptocurrency, but instead of a decentralized finance (DeFi) coin, you opt for a social media cryptocurrency. Since it's a social media-based coin and not a DeFi one, you feel at ease knowing that the company hasn't registered its cryptocurrency with the SEC.

LBRY is a blockchain-based social media video-sharing platform. In U.S. Securities and Exchange Commission v. LBRY (SEC v. LBRY), the SEC alleged LBRY sold $11 million in LBRY credits (LBC) as securities; however, LBRY did not register LBC as a security with the SEC, and LBRY credits did not qualify for a securities exemption. LBRY argued that LBCs were primarily utility tokens used within the LBRY platform to access features like content creation and tipping creators. LBRY argued that LBCs were not securities because they did not conduct an ICO and that LBC does not meet the Howey Test criteria, particularly with respect to dependence on their efforts to generate profits.[95]

In November 2022, the court ruled in favor of the SEC. The judge concluded that LBCs were unregistered securities based on the Howey Test.

The SEC v. LBRY ruling is significant because LBRY is a decentralized social media company, not a decentralized finance (DeFi) cryptocurrency like Bitcoin or a centralized finance (CeFi) cryptocurrency like XRP. As a blockchain-decentralized social media platform, most investors did not consider LBRY a security.

---

[95] Reiners, Lee. How We Got Here and How We Get Out – Defining Security Status. SEC.gov, March 21, 2025. https://www.sec.gov/files/ctf-input-reiners-2025-3-18.pdf (Accessed October 27, 2025)

LESSON LEARNED: Generally, if a cryptocurrency company claims that its cryptocurrency can increase in value, it may be considered a security and should register as such with the SEC. Investors are encouraged to use SEC.gov to determine if a cryptocurrency is registered as a security before committing any investment dollars. According to a conventional estimate, there are potentially thousands of cryptocurrencies and tokens in use that should probably be registered with the SEC or other global securities regulatory agencies as securities.

As a footnote to BlockFi, NASGO, and LBRY, I knew about all three of these cryptocurrency companies in their early days. Still, I decided not to invest because they did not register their cryptocurrencies or crypto-lending products with the SEC.

The next section will address legal issues related to the SEC and cryptocurrency exchanges.

CRYPTOCURRENCY EXCHANGE LEGAL ISSUES

Before depositing money into a cryptocurrency exchange, research the exchange to ensure it is compliant with applicable laws and regulations. This is especially true if you are depositing a large amount of money. The next story will demonstrate how recognizing legal issues and asking the right questions can help you avoid a poor investment decision.

In the fall of 2020, I received an email from the owner of a European cryptocurrency exchange, inquiring about my interest in purchasing the exchange. The exchange seemed legitimate, as it traded "blue chip" cryptocurrencies, such as Bitcoin, Ethereum, and Litecoin. I thought it would be cool to own a cryptocurrency exchange, so I emailed him with three legal questions:

1. Is the cryptocurrency exchange registered with the SEC to operate in the US?
2. Does the cryptocurrency exchange have enough business insurance to cover hacks and other IT threats to replace investors' money if it were stolen?
3. Do you have a cryptocurrency exchange user agreement you can send me to review?

The answer? He never responded to my email.

My instinctive reaction was that he might have another buyer. However, I quickly realized that if I was willing to make an offer, he would have a bidding war; and if he had two or more bidders, he'd likely receive his asking price. I expect he did not respond because he either did not register the exchange with the SEC, did not have insurance to cover losses from hacking and other IT threats, or did not want me to see the cryptocurrency exchange user agreement. For any (or all) of these reasons, I determined the investment was not worth the risk.

Since most people use exchanges to buy and sell cryptocurrencies, it is essential to understand the type of cryptocurrency exchange you are using, the kind of user agreement you are signing, and the legal issues you may face when using a cryptocurrency exchange. This becomes particularly necessary considering that numerous cryptocurrency exchanges, including FTX, Voyager, and Celsius, all filed for bankruptcy in 2022, resulting in billions of dollars being lost by exchange users.[96]

---

[96] Llamas, Michele. Crypto Bankruptcies. ConsumerNotice.org, March 22, 2024. https://www.consumernotice.org/legal/crypto-bankruptcies/ (September 3, 2025)

Even if you are not in the legal, legislative, or regulatory field, it is important to understand legal issues related to cryptocurrency exchanges. Some of the most common cryptocurrency exchange legal issues include:

- Is the cryptocurrency exchange in compliance with government regulations? Cryptocurrency exchanges are subject to laws and regulations that vary widely from one jurisdiction to another. For example, in the US, cryptocurrency exchanges are heavily regulated and must be registered with the SEC, whereas in other countries, they may operate in a largely unregulated environment.
- Does the cryptocurrency exchange have adequate security measures to protect users' cryptocurrency? Hackers frequently target cryptocurrency exchanges since some exchanges transact billions of dollars' worth of cryptocurrency. If the exchange is not properly secured, it can be vulnerable to hacks and cyber-attacks, resulting in the loss of funds for exchange users and legal liability.
- Does the cryptocurrency exchange respond to customer disputes? Cryptocurrency exchanges may receive complaints from customers about issues such as lost or stolen funds, technical glitches, or disputes over the terms of service. The exchange's response to customer complaints is usually based on the user agreement, which is issued by the exchange and virtually signed by the user. These disputes can result in mediation, arbitration, or legal action against the exchange.

Truthful and transparent answers to these questions can be found on government websites (SEC, CFTC, and Treasury), social media platforms (e.g., Twitter, Discord, and Telegram), and websites that rate and review cryptocurrency exchanges (e.g., Certified - Cer.live).

Finally, cryptocurrency exchanges should establish customer service departments that actually respond to and resolve customer complaints. I know many cryptocurrency investors who had their wallets hacked. Despite repeated attempts to contact the cryptocurrency exchange, these investors received standardized emails dismissing responsibility and liability for returning the funds. To safeguard your investments, conduct thorough research and select a cryptocurrency exchange that has transparent customer policies or a proven track record of addressing user concerns. The more transparent the exchange is, the safer your investment will be!

SEC ENFORCEMENT ACTIONS AGAINST CRYPTOCURRENCY EXCHANGES

In 2018, the SEC also started investigating and conducting enforcement actions against cryptocurrency exchanges operating in the US. EtherDelta was the first cryptocurrency exchange the SEC investigated, resulting in a settlement of $300,000.00, along with interest and penalty payments. [97]This section of the chapter will cover the SEC's enforcement of cryptocurrency exchanges.

BINANCE HOLDINGS LTD.

In 2017, Changpeng Zhao founded Binance Holdings Ltd. (Binance) in China. Binance is the largest global cryptocurrency exchange, with an average daily volume of $65 billion. In 2019, the US government banned Binance. In response, Binance created BAM Trading Services, the operating company for Binance US, a cryptocurrency exchange, but did not register the latter as a cryptocurrency exchange with the SEC.[98]

In 2023, the SEC sued Binance in the US District Court for the District of Columbia for operating as an unregistered securities exchange. They

---

[97] Bennett, Jennifer. SEC Settles First Unregistered Crypto Securities Exchange Case. News.Bloookberglaw.com, November 8, 2018. https://news.bloomberglaw.com/securities-law/sec-settles-first-unregistered-crypto-securities-exchange-case-1?context=article-related (Accessed September 3, 2025)

[98] Press Release. SEC Files 13 Charges Against Binance Entities and Founder Changpeng Zhao. SEC.gov, June 5, 2023. https://www.sec.gov/newsroom/press-releases/2023-101 (Accessed September 3, 2025)1

alleged that Binance violated critical registration-related provisions of the federal securities laws. The SEC filed charges against Binance and BAM Trading Services, including:

- Operating unregistered national securities exchanges, broker-dealers, and clearing agencies;
- The unregistered offer and sale of Binance's crypto assets, including: BNB, an exchange token, and Binance USD (BUSD), a so-called stablecoin, certain crypto-lending products, and a staking-as-a-service program.[99]

In November 2023, Chagpeng Zhao pleaded guilty to failing to maintain an effective anti-money laundering (AML) program, failing to register as a money-transmitting business, and violating the International Emergency Economic Powers Act (IEEPA) under the BSA. Zhao was sentenced to four months in prison, resigned as Binance's CEO, and agreed to pay over $4 billion to resolve the Justice Department's investigation into violations of the Bank Secrecy Act (BSA).[100] As an endnote to the Binance legal saga, in 2025, President Trump pardoned Changpeng Zhao.[101]

POLOINEX

Poloinex is a cryptocurrency exchange that supports over 400 cryptocurrencies, including major coins such as Bitcoin (BTC), Ethereum (ETH), Ripple (XRP), and numerous altcoins. It also offers Poloinex Futures and an NFT trading platform. Although Poloinex sold securities to

---

[99] Press Release. SEC Files 13 Charges Against Binance Entities and Founder Changpeng Zhao. SEC.gov, June 5, 2023. https://www.sec.gov/news/press-release/2023-101 (Accessed October 27, 2025)

[100] Binance and CEO Plead Guilty to Federal; Charges in $4 Billion Resolution. Justice.gov, November 21, 2023. https://www.justice.gov/opa/pr/binance-and-ceo-plead-guilty-federal-charges-4b-resolution (Accessed October 27, 2025)

[101] Treene, Alayna and Morrow, Allison. Trump Pardons Binance Founder Changpeng Zhao. Cnn.com, October 23, 2025. https://www.cnn.com/2025/10/23/politics/binance-founder-trump-pardon (Accessed October 27, 2025)

US investors on its platform, it did not register with the SEC as a securities exchange filing or for an exemption. The SEC found that Poloinex met the definition of an "exchange" under Section 5 of the Exchange Act because Poloinex "provided the non-discretionary means for trade orders to interact and execute through the combined use of the Poloinex, an order book, and the Poloinex trading engine."[102]

Without admitting or denying any wrongdoing, Poloinex settled with the SEC and paid more than $10 million for "operating an unregistered online digital asset exchange in connection with its operation of a trading platform that facilitated the buying and selling of digital asset securities." [103]

COINBASE

Founded by Brian Armstrong in 2012, at the time of publication, Coinbase was the number 2 cryptocurrency exchange by trading volume. In SEC v. Coinbase Inc. and Coinbase Global Inc. (SEC v. Coinbase), the SEC sued Coinbase in the SDNY Court for violating federal securities laws by operating as an unregistered broker, exchange, and clearing agency for cryptocurrencies that were securities.[104]

According to the SEC's complaint, since at least 2019, Coinbase has generated billions of dollars by unlawfully facilitating the purchase and sale of cryptocurrencies as securities. The SEC alleged that Coinbase:

---

[102] U.S. Securities and Exchange Commission. "In the Matter of Poloniex, LLC Admin. Proc. File No. 3-20455." U.S. Securities and Exchange Commission, March 3, 2023. https://www.sec.gov/enforcement/information-for-harmed-investors/poloniex. (Accessed October 27, 2025)

[103] Ibid.

[104] De, Nikhiulesh. "Coinbase Moves to Dismiss SEC Lawsuit , Alleging Crypto Falls Out of Regulator's Oversight." Coindesk, August 4, 2023. https://www.coindesk.com/policy/2023/08/04/coinbase-moves-to-dismiss-sec-lawsuit-alleging-crypto-falls-out-of-regulators-oversight/#:~:text=The%20SEC%20sued%20Coinbase%20in,it%20believed%20met%20those%20requirements (Accessed September 3, 2025)

- Provides a marketplace and brings together the orders for securities of multiple buyers and sellers using established, non-discretionary methods under which such orders interact;
- Engages in the business of effecting securities transactions for the accounts of Coinbase customers; and
- Provides facilities for comparing data regarding the terms of settlement for crypto asset securities transactions, serves as an intermediary in settling transactions in crypto asset securities for Coinbase customers, and acts as a securities depository.[105]

In response, Coinbase argued that regulating the cryptocurrency industry falls beyond the SEC's jurisdiction.

In February 2025, under the Trump Administration, the SEC dismissed the case against Coinbase, stating the following:

> On January 21, 2025, the Commission announced the formation of the Crypto Task Force, which is dedicated to helping develop a comprehensive and clear regulatory framework for crypto assets. Given the pending work of the Crypto Task Force, the Commission is dismissing this matter.[106]

The SEC also stated the Cyber and Emerging Technologies Unit will continue to "root out those seeking to misuse innovation to harm investors, including fraud involving blockchain technology and crypto assets".

---

[105] U.S. Securities and Exchange Commission. "SEC Charges Coinbase for Operating as an Unregistered Securities Exchange, Broker, and Clearing Agency." U.S. Securities and Exchange Commission, June 6, 2023. https://www.sec.gov/news/press-release/2023-102. (Accessed September 3, 2025)

[106] Press Release. SEC Announces Dismissal of Civil Enforcement Action Against Coinbase. SEC.gov, February 27, 2025. https://www.sec.gov/newsroom/press-releases/2025-47 (Accessed November 1, 2025)

COMMODITIES AND FUTURES TRADING COMMISSION

In recent years, the Commodities and Futures Trading Commission (CFTC) has become more active in regulating cryptocurrency companies. Moreover, the CFTC ruled that Bitcoin is a commodity. *One question remains*: what about altcoins? Are altcoins commodities as well?

The answer is: it depends.

There's an ongoing discussion about whether cryptocurrencies should be seen as securities or commodities. Some regulators and market participants consider cryptocurrencies commodities, while others believe they possess special features that distinguish them from traditional commodities. Here's a straightforward and clear summary of the main arguments for and against viewing cryptocurrencies as commodities.

Arguments for Cryptocurrencies as Commodities:

- **Store of Value**: Like commodities, cryptocurrencies can be used as a store of value.
- **Price Fluctuations**: Both commodities and cryptocurrencies are subject to price fluctuations based on market demand and supply.
- **Speculative Trading**: Both commodities and cryptocurrencies are often traded for speculative purposes.
- **Scarcity:** Cryptocurrencies with a finite total supply are more similar to commodities, as they inherently possess an overall scarcity component.

Arguments Against Cryptocurrencies as Commodities:

- **Digital Nature**: Cryptocurrencies are digital assets, unlike physical commodities, like gold, silver, corn, and soybeans.
- **Unique Functions**: Some cryptocurrencies have unique functions, like smart contracts and decentralized applications, that are more akin to business transactions than traditional commodities.

Ultimately, the classification of cryptocurrencies as commodities can vary depending on the jurisdiction and the specific cryptocurrency in question. The ongoing debate over altcoin commodities highlights the complexity and evolving nature of the cryptocurrency market.

CFTC INVESTIGATES BITMEX IN TWO CASES

SCENARIO: You want to invest in cryptocurrencies, so you find an exchange that trades cryptocurrencies and derivatives. You want to make sure that the cryptocurrency exchange is in compliance, so you research whether it is registered with the SEC. Should you research if the exchange is registered with any other government agencies?

THE CFTC INVESTIGATES BITMEX

Founded in 2014 and registered in Seychelles, BitMEX is a cryptocurrency exchange and derivatives trading platform. The CFTC charged BitMEX's owners with operating an unregistered cryptocurrency derivatives platform and violating anti-money laundering regulations.

In the first BitMEX case, Commodities Futures Trading Commission v. BitMEX (CFTC v. BitMEX), the US District Court for the Southern District of New York court (SDNY Court) found that virtual currencies fell within the catch-all portion of CEA's definition of "commodity," which includes, "all other goods and articles . . . and all services, rights and interests in which contracts for future delivery are presently or in the future dealt in." 7 U.S.C. § 1(a)(9).The Federal Court ruled in favor of the CFTC and fined BitMEX $30 million.[107]

---

[107] CFTC. Federal Court Orders BitMEX's Three Co Founders to Pay a Total of $30 Million for Illegally Operating a Cryptocurrency Derivatives Trading Platform and Anti-Money Laundering Violations. CFTC.gov, May 5, 2022. https://www.cftc.gov/PressRoom/PressReleases/8522-22 (Accessed September 3, 2025)

In the second BitMEX case, the United States of America v. Arthur Hayes, Benjamin Delo, Samuel Reed, and Gregory Dwyer (US v. BitMEX) in the SDNY, the US Attorney alleged that BitMEX violated the Bank Secrecy Act because it did not comply with the Know Your Customer federal law, which resulted in BitMEX being used as a vehicle for money laundering. The two BitMEX founders pleaded guilty to violating the Bank Secrecy Act and agreed to pay $10 million each as part of a plea deal.[108]

Most investors only pay attention to SEC registration and enforcement for cryptocurrencies. Still, the BitMEX cases demonstrate that cryptocurrency exchange companies may also need to register and comply with CFTC regulations.

LESSON LEARNED 1: Be cautious when using offshore cryptocurrency exchanges or those outside your country. If anything goes wrong, the regulators in your country may not be able to enforce their laws on a cryptocurrency exchange outside of their jurisdiction.

LESSON LEARNED 2: Research the cryptocurrency exchange's reputation and ensure it is registered and compliant with your country's laws and regulations. If the exchange is not compliant, regulators may shut down the exchange, and you may lose some or all of your investment.

LESSON LEARNED 3: Diversify your cryptocurrency holdings in different exchanges. Using multiple cryptocurrency exchanges will protect your overall cryptocurrency investment in case one of the exchanges faces regulatory issues or goes out of business. You can start with small transactions to test the platform's reliability and ease of use.

---

[108] Cohen, Luc and Godoy, Jody. "Founders of Crypto Exchange BitMEX Plead Guilty to Bank Secrecy Act Violations." Reuters.com, February 22, 2022. https://www.reuters.com/technology/founders-crypto-exchange-bitmex-plead-guilty-bank-secrecy-act-violations-2022-02-24/. (Accessed September 19, 2025)

CFTC INVESTIGATES TETHER

SCENARIO: Going by the word "stable" in stablecoins, you think they are less risky than cryptocurrencies. So, you transfer your cryptocurrency profits to the first stablecoin you find until you can decide which cryptocurrency you want to invest in, thinking you're being cautious and making a safe investment.

In the matter of Tether Holdings Limited, the CFTC alleged that Tether made untrue or misleading statements and omitted material facts in connection with the Tether (USDT) stablecoin. Specifically, the CFTC alleged that Tether misrepresented to customers and the market that they maintained sufficient U.S. dollar reserves to back every USDT in circulation with the "equivalent amount of corresponding fiat currency" held by Tether and "safely deposited" in Tether's bank accounts.[109]

Tether Holdings Limited settled with the CFTC, agreeing to pay a $42 million fine for "making untrue or misleading statements and omissions" about the backing of its stablecoin tokens.

LESSON LEARNED 1: Even though stablecoins are designed to be "stable" and less risky investments, thoroughly research the stablecoin before you invest. Some stablecoins may not be transparent about the assets backing their value. A lack of thorough audits can hide financial instability or fraud. As with all cryptocurrencies, you are advised to thoroughly research stablecoins before you commit your investment dollars.

LESSON LEARNED 2: There are numerous types of stablecoins backed by different types of collateral. If a stablecoin is pegged to a currency

[109] Reyonlds, Sam. Tether Settles with CFTC for $42.5M. Blockworks.co, October 15, 2021. https://blockworks.co/news/tether-settles-with-cftc-for-42-5m (Accessed October 27, 2025)

experiencing high inflation, its purchasing power may decrease over time. In this scenario, the stablecoin may not be a very "stable" investment.

THE DEPARTMENT OF JUSTICE

Since 2019, the Department of Justice has increased the criminal enforcement of cryptocurrency companies that violate criminal laws. The Department of Justice's mission is "to uphold the rule of law, keep our country safe, and protect civil rights."[110]

The following investigations and cases show the Department of Justice's (DOJ) cryptocurrency enforcement efforts.

SCENARIO: Many of your friends invested in cryptocurrencies in 2018 and made a good return on their investment. You now regret not following suit then, and wish to start investing in cryptocurrencies. So you start researching cryptocurrency investment opportunities and even find a promising cryptocurrency. You are now drawn to the statement that cryptocurrency offers "guaranteed returns." You invest because they state your investment is "guaranteed". This scenario applies to the next two DOJ investigations.

$100 MILLION CRYPTOCURRENCY PONZI SCHEME

Ponzi schemes, named after the infamous swindler Charles Ponzi, present themselves as legitimate investments that are, in fact, scams. Ponzi scheme organizers attract new investors by falsely promising to invest funds in opportunities that will generate high returns with little or no risk to the investor. Instead of investing that money in a legitimate opportunity, the organizers use new investor money to make promised payments to existing investors while keeping the "invested funds" for personal use. The last

---

[110] Our Work. Justice.gov. https://www.justice.gov/our-work (Accessed October 27, 2025)

investors lose their money, because the Ponzi scheme no longer attracts investors to "invest new funds."

In U.S. v. Emerson Pires, et al., the DOJ alleged that three men used their company, EmpiresX, to run a cryptocurrency Ponzi scheme that raised $100 million. EmpiresX offered "guaranteed returns" to investors. The DOJ accused EmpiresX of misappropriating most of the money and using funds from new investors to pay current investors, a typical practice in Ponzi schemes. The DOJ reports that the head trader pleaded guilty to one count of securities fraud.[111]

$2 BILLION CRYPTOCURRENCY FRAUD

In U.S. v. Arcaro, the DOJ charged BitConnect with defrauding investors through a pyramid scheme known as the BitConnect Referral Program. BitConnect persuaded investors to "get in early" and promised "guaranteed returns," using social media videos and posts to defraud thousands of investors of over $2 billion.[112]

LESSON LEARNED: Any investments, including cryptocurrencies, that offer "guaranteed returns" are probably scams. Legitimate securities investments involve risks and cannot guarantee a rate of return.

CRYPTOCURRENCY MIXER BASICS

Cryptocurrency mixers are privacy-enhancing tools used to increase the anonymity of cryptocurrency transactions. They work by combining funds

---

[111] U.S. Department of Justice. "Three Men Charged in $100 Million Cryptocurrency Fraud." U.S. Department of Justice, June 30, 2022. https://www.justice.gov/usao-sdfl/pr/three-men-charged-100-million-cryptocurrency-fraud. (Accessed October 27, 2025)

[112] U.S. Department of Justice. "Director and Promoter of BitConnect Pleads Guilty on Global $2 Billion Cryptocurrency Scheme." U.S. Department of Justice, September 1, 2021. https://www.justice.gov/usao-sdca/pr/director-and-promoter-bitconnect-pleads-guilty-global-2-billion-cryptocurrency-scheme. (Accessed September 3, 2025)

from multiple users and then redistributing them, making it difficult to trace their source.[113]

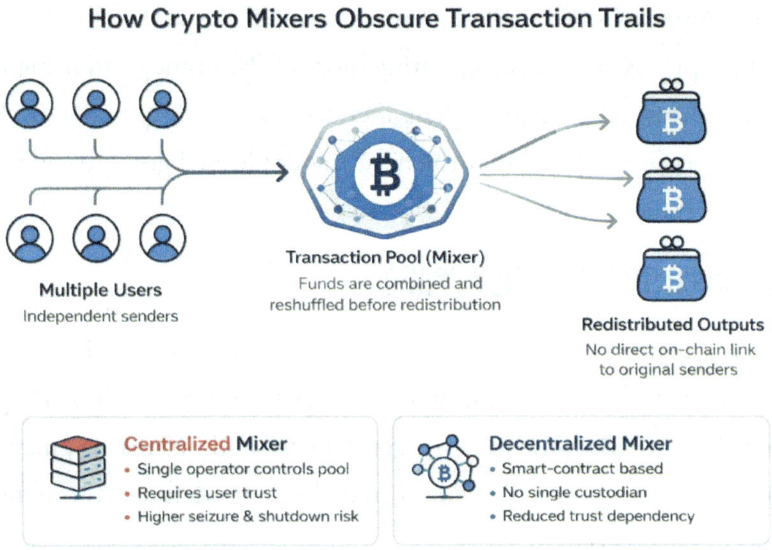

Cryptocurrency mixers, also known as tumblers, typically use the following elements to increase anonymity in their transactions:

- **Pooling Funds**: Users send their cryptocurrencies to the mixer, which combines these funds into a large reserve.
- **Breaking the Link**: Then, the mixer sends the same amount of cryptocurrency (minus a service fee) to different wallet addresses, often in smaller and varied amounts. This process breaks the link between the sender and receiver, making it challenging to trace the origin of the funds.

---

[113] Arora, SK. What is a Cryptocurrency Mixer and How Does it Work? Cointelegraph.com, May 27, 2025. https://cointelegraph.com/explained/what-is-a-cryptocurrency-mixer-and-how-does-it-work (Accessed October 27, 2025)

- **Enhancing Privacy**: By concealing the transaction trail, mixers allegedly increase privacy and anonymity for users who want to hide their cryptocurrency transactions.

There are two types of mixers: centralized and decentralized. Centralized mixers offer a simpler solution, but they raise privacy concerns, while decentralized mixers provide a higher level of privacy, albeit at the cost of greater complexity. The choice between the two depends on your needs and your risk tolerance.

CENTRALIZED CRYPTOCURRENCY MIXERS

Centralized cryptocurrency mixers are run by a single entity, usually a company. Although they might offer some privacy, there's always a chance that the operator could retain transaction data or cooperate with law enforcement. Centralized mixers have also been linked to scams, such as exit scams, where the operator walks away with the funds.

DECENTRALIZED CRYPTOCURRENCY MIXERS

Similar to blockchain technology, decentralized mixers operate on a distributed network of nodes. Because there's no single point of control, it is much harder for anyone to track or trace transactions. The decentralized nature of these mixers makes them less vulnerable to attacks and scams. However, decentralized mixers can be more complex to set up and use and may require specific technical knowledge.

REGULATION OF CRYPTOCURRENCY MIXERS

While mixers can offer increased privacy, they have been linked to illegal activities, including money laundering, funding criminal activities, and facilitating illicit transactions. As a result, the use of mixers raises concerns with regulatory bodies, and some jurisdictions have imposed restrictions or regulations on their operation. For example, FinCEN has defined mixers as money transmitters. In compliance with the Bank Secrecy Act, they must meet three key obligations:

1. Register with FinCEN.
2. Maintain an anti-money laundering and know-your-customer (KYC) compliance program.
3. Meet all applicable reporting and record-keeping requirements.

However, remember that using mixers does not guarantee complete anonymity, and their effectiveness depends on various factors, including the service's trustworthiness, the specific blockchain's transparency, and potential vulnerabilities in the mixing process.

DARKNET CRYPTOCURRENCY MIXER OWNERS INDICTED

In U.S. v. Sterlingov, the DOJ arrested and indicted the owner of Bitcoin Fog, which the DOJ alleges is a darknet cryptocurrency mixer.[114] The BitcoinFog.info site describes their services:

> You register an account on our website and deposit bitcoins to the designated address. After a couple of network confirmations, your money is registered on your Fog account, and you can schedule withdrawals. Each withdrawal will be split into a random number of payouts, depending on the amount, and the relative size of each payout will also be randomized. Even the timing of those payouts will be randomly spread out over the period you specify.[115]

---

[114] U.S. Department of Justice. "Individual Arrested and Charged with Operating Notorious Darknet Cryptocurrency Mixer." U.S. Department of Justice, April 28, 2021. https://www.justice.gov/opa/pr/individual-arrested-and-charged-operating-notorious-darknet-cryptocurrency-mixer. (Accessed September 3, 2025)

[115] Accessing Bitcoin Fog. BitcoinFog.Info https://bitcoinfog.info. (Accessed October 28, 2025)

Did you notice how the word "*random*" seems to appear a bit too frequently in the above description? Major red flag!

Some of the terms that indicate a cryptocurrency is actually a cryptocurrency mixer are:

- "Random number of payouts"
- "Randomized"
- "Randomly spread"

If you see terms like these in a description of a cryptocurrency, beware — it's probably a cryptocurrency mixer. Similar to Tornado Cash, another alleged cryptocurrency mixer, the U.S. Government is tracking and targeting cryptocurrency mixers as they may be used to launder money.

THE BIG PICTURE

Cryptocurrency mixers have a very uncertain future. I predict that more global regulators will increase their regulations of cryptocurrency mixers to prevent money laundering, funding criminal activity, and facilitating illicit transactions. The future of cryptocurrency mixers will likely depend on finding a balance between privacy and security. While there is a legitimate need for privacy in financial transactions, it is crucial to prevent their misuse for criminal purposes.

The next section will analyze the Department of the Treasury's role in cryptocurrency enforcement.

THE DEPARTMENT OF THE TREASURY

The Department of the Treasury (Treasury) is another US agency that has increased regulation and enforcement of cryptocurrency companies.

Treasury's mission is to maintain a strong economy and create economic and job opportunities by promoting the conditions that enable economic growth and stability at home and abroad, strengthen national security by combating threats and protecting the integrity of the financial system, and manage the U.S. Government's finances and resources effectively.

Let's see how the role of the Treasury plays out in the next scenario.

SCENARIO: You are invited to a meeting to invest in a cryptocurrency exchange. You want to make sure that the exchange you're potentially investing in is compliant with federal regulations, so you ask if it is registered with the SEC. The representative for the cryptocurrency exchange verifies that it is registered with the SEC and fully compliant with all government regulations. You are encouraged because the exchange is compliant and offers a good investment opportunity, right? Wrong! There are a few important follow-up questions you must ask. The next section covers the answer to this scenario.[116]

THE IMPORTANCE OF THE BANK SECRECY ACT

The Bank Secrecy Act of 1970 (BSA), also known as the Currency and Foreign Transactions Reporting Act, requires US financial institutions to assist US government agencies in detecting and preventing money laundering. Originally, the BSA applied to banks and credit unions. However, as the financial industry diversified to include new types of businesses, such as cryptocurrency businesses, these businesses are also covered under the BSA and Anti-Money Laundering (AML) regulations.[117]

The BSA has been amended several times, including provisions in Title III of the USA PATRIOT Act. Section 326 of the USA PATRIOT Act

---

[116] Role of the Treasury. https://home.treasury.gov/about/general-information/role-of-the-treasury (Accessed September 3, 2025)

[117] Lemire, Katherine. Cryptocurrency and Anti-Money Laundering Enforcement. Reuters.com, September 26, 2022. https://www.reuters.com/legal/transactional/cryptocurrency-anti-money-laundering-enforcement-2022-09-26/ (Accessed October 28, 2025)

requires financial institutions to develop and implement a Customer Identification Program (CIP) that verifies customer identity, compares names with terrorist lists, and maintains appropriate recordkeeping. In addition, the USA PATRIOT Act requires all financial institutions, including money services businesses (MSBs), to comply with all the BSA and AML requirements. For example, cryptocurrency exchanges, which trade fiat currencies for cryptocurrencies or vice versa, are considered MSBs.[118]

Finally, the BSA requires financial institutions to have a BSA/AML compliance program. Think of the BSA as a way to shine a light on money that might be used for illegal purposes, like money laundering or funding terrorism. This helps law enforcement agencies illicit gains and protect the integrity of the US financial system.

THE FINANCIAL CRIMES NETWORK AND OFFICE OF FOREIGN ASSETS CONTROL

The Treasury utilizes the Financial Crimes Enforcement Network (FINCEN) and the Office of Foreign Assets Control (OFAC) divisions to investigate and enforce the improper use of cryptocurrencies. According to the OFAC's website, its mission is to administer and enforce "economic and trade sanctions based on US foreign policy and national security goals against targeted foreign countries and regimes, terrorists, international narcotics traffickers, those engaged in activities related to the proliferation of weapons of mass destruction, and other threats to the national security, foreign policy or economy of the United States."[119]

---

[118] From the Examiners Desk. FDIC.gov.
https://www.fdic.gov/regulations/examinations/supervisory/insights/sisum04/sisummer04-article5.pdf
(Accessed October 27, 2025)

[119] Mission. Office of Foreign Assets Controls. OFAC.Treasury.gov.
https://ofac.treasury.gov/#:~:text=The%20Office%20of%20Foreign%20Assets,traffickers%2C%20those%20engaged%20in%20activities (Accessed October 28, 2025)

Financial institutions are required to file confidential documents, known as Suspicious Activity Reports (SARs), with government authorities when they detect suspicious or potentially illegal activities. SARs are a crucial tool in the fight against financial crimes, such as money laundering, fraud, and terrorist financing. By reporting suspicious activity, financial institutions can help authorities identify potential criminal activity in its early stages, leading to quicker investigations and potentially preventing crimes. The information gathered from SARs helps government agencies, such as the US Financial Crimes Enforcement Network (FinCEN), identify trends and patterns associated with financial crimes.

The next section will cover three Treasury enforcement actions of well-known companies in the cryptocurrency space.

BITTREX SANCTIONED NEARLY $30 MILLION

Bittrex is a cryptocurrency exchange that started in 2014. In 2021, FINCEN and OFAC conducted independent investigations into Bittrex. They found that Bittrex had failed to file Suspicious Activity Reports (SARs), had insufficient anti-money laundering efforts, failed in its sanctions efforts, and had links to darknet markets, mixers, and ransomware gangs. These are four very important and troubling findings.

OFAC INVESTIGATION

The OFAC investigation also found that Bittrex had deficient sanctions compliance procedures and failed to prevent persons from sanctioned jurisdictions, including the Crimea region of Ukraine, Cuba, Iran, Sudan, and Syria, from using the Bittrex platform. As part of the investigation, OFAC considered the following mitigating factors:

1. Bittrex was a small and new company.

2. Bittrex provided "substantial cooperation" with the OFAC investigation.
3. Most of the investigated transactions were for a "relatively small amount" and were a "relatively small percentage" of Bittrex's annual transaction volume.
4. Bittrex took swift remedial measures to curtail the violations.[120]

These mitigating factors likely reduced the sanctions that OFAC could have levied against Bittrex.

FINCEN INVESTIGATION

The FinCEN investigation found that Bittrex did not maintain an effective AML program from February 2014 through December 2018, essentially the first four years of its operation. Treasury reports that Bittrex settled the case for $24 million with OFAC and $29 million with FinCEN.[121]

ANALYSIS OF THE FINCEN AND OFAC FINDINGS

The FinCEN investigation also underscores the importance of being aware of and adhering to international sanctions. The importance of awareness extends not only to cryptocurrency companies but also to investors. As an individual investor, it is essential to stay informed about relevant regulations, particularly when dealing with cryptocurrencies associated with sanctioned countries or entities.

The Bittrex regulatory story serves as a cautionary tale, highlighting the importance of a well-resourced compliance department in cryptocurrency

---

120 Department of Treasury. "OFAC Settles with Bittrex, Inc. for $24,280,829.20 Related to Apparent Violations of Multiple Sanctions Programs." U.S. Department of Treasury, October 11, 2022. https://ofac.treasury.gov/media/928746/download?inline#:~:text=As%20a%20result%20of%20deficiencies,of%20virtual%20currency%2Drelated%20transactions. (Accessed October 28, 2025)

121 Department of Treasury. "Treasury Announces Two Enforcement Actions for over $24 M and $29M Against Virtual Currency Exchange Bittrex, Inc." October 11, 2022. https://home.treasury.gov/news/press-releases/jy1006. (Accessed September 3, 2025).

companies and exchanges, even for small or newly established entities. Before deciding to invest, carefully review the cryptocurrency exchange's website for detailed information about its compliance program.

You reduce the risk of having issues with your cryptocurrency exchange when you invest in exchanges that are registered with government regulators. Regulated exchanges typically have stricter compliance measures in place to prevent violations of sanctions and other laws. In the US, it is best practice to research the cryptocurrency exchange in the OFAC Civil Penalties and Enforcement Information Section to determine if the exchange has any compliance issues. [122]

During my discussions with companies seeking legal advice, I have observed that many are reluctant to invest time and resources in achieving full compliance with the US government agencies. Strong and well-resourced compliance departments should be an integral part of all cryptocurrency companies. Potential investors must ensure that they ask about the company's US and global compliance efforts. Although the Treasury investigation did not find that Bittrex intentionally engaged in illegal activities, if Bittrex had established a well-resourced compliance department at inception, it may not have been sanctioned by the Treasury.

TORNADO CASH SANCTIONED BY OFAC

SCENARIO: You use a cryptocurrency mixer because you believe cryptocurrency transactions should be anonymous, and you do not want anyone to trace your transactions. Since you are not doing anything illegal with your cryptocurrency, you think nothing of using a mixer — you're doing nothing wrong, after all.

---

[122] Civil Penalties and Enforcement Information. Office of Foreign Assets Controls. https://ofac.treasury.gov/civil-penalties-and-enforcement-information (Accessed October 28, 2025)

Founded in 2019, Tornado Cash is a virtual cryptocurrency mixer with a cryptocurrency (TORN) on the Ethereum blockchain. TORN uses smart contracts to disrupt the on-chain link between the sender and recipient addresses. The on-chain disruption makes it more difficult to track cryptocurrency transactions, but not impossible. In August 2022, OFAC officially sanctioned Tornado Cash for laundering $7 billion in digital assets since 2019.[123]

Tornado Cash is one of Treasury's most important enforcement actions at the time of publication. The OFAC sanctions had the following impact on Tornado Cash:

- A Tornado Cash developer was arrested.
- All Tornado Cash property and interests in the US, or in possession or control of US persons, were blocked and must be reported to OFAC.[124]

In 2022, Brian E. Nelson, Under Secretary of the Treasury for Terrorism and Financial Intelligence, made the following statement regarding the Tornado Cash OFAC sanction:

> "Treasury is sanctioning Tornado Cash, a virtual currency mixer that launders the proceeds of cybercrimes, including those committed against victims in the United States. Despite public assurances to the contrary, Tornado Cash has repeatedly failed to implement effective controls designed to prevent it from laundering funds for malicious cyber actors on a regular basis, and without taking basic measures to address its risks.

---

[123] US Department of Treasury. "U.S. Treasury Sanctions Notorious Virtual Currency Mixer Tornado Cash." U.S. Department of Treasury, August 8, 2022, https://home.treasury.gov/news/press-releases/jy0916. (Accessed September 3, 2025)

[124] Ibid.

Treasury will continue to aggressively pursue actions against mixers that launder virtual currency for criminals and those who assist them."[125]

OFAC's enforcement action against Tornado Cash and Secretary Nelson's comments clearly indicate that OFAC intends to pursue cryptocurrency mixers and perhaps even privacy coins.

I've often had insightful conversations with fellow attorneys and cryptocurrency experts about how the OFAC sanctions on Tornado Cash have deeply affected the cryptocurrency community. We've especially focused on issues related to privacy, DeFi, and the overall changes in the regulatory landscape. Here are some key points that frequently come up in these discussions:

1. The Tornado Cash sanctions raised concerns about the erosion of privacy in the cryptocurrency space. Users who had interacted with Tornado Cash faced potential legal risks, and developers became hesitant to create similar tools.
2. Tornado Cash was a crucial component of many decentralized applications (dApps), so the sanctions had a direct impact on the decentralized finance (DeFi) ecosystem. Its removal led to a decline in the popularity of certain DeFi protocols that relied on Tornado Cash for privacy features.
3. The sanctions marked a significant step by the U.S. government in regulating the cryptocurrency space because they targeted a cryptocurrency mixer that could be used for illicit activities.
4. The sanctions sparked a debate within the cryptocurrency community about the balance between privacy and security. Some argued that the sanctions were necessary to prevent money laundering and other illicit

[125] Ibid.

activities, while others criticized them as an attack on privacy and individual freedoms. Of course, there was a third camp — which included me — advocating for a balance between the two.

The long-term implications of the Tornado Cash sanctions are still unfolding. While these sanctions may have temporarily dampened innovation in certain areas, they also highlighted the need to develop more robust privacy solutions that comply with regulatory requirements.

LESSON LEARNED: Cryptocurrency mixers and privacy coins face significant KYC/AML compliance issues, resulting in more regulatory uncertainty compared to regular cryptocurrencies. Investors should not overlook the risk of heightened regulatory uncertainty associated with cryptocurrency mixers or privacy coins. Here's a simple rule: increased regulatory uncertainty means increased risk in investment!

KRAKEN SETTLES WITH OFAC FOR $362,158.70

SCENARIO: You create a decentralized finance cryptocurrency and raise $3 million to fund your business for three years. Your excitement knows no bounds — your cryptocurrency will be listed on major cryptocurrency exchanges, and anyone in the world can download your wallet and buy it. You file the appropriate regulations with the SEC and CFTC. You might think you're ready to release a new, groundbreaking cryptocurrency, but you need to consider a few other regulatory filings.

OFAC's investigation into Kraken is another example of the importance of a well-resourced compliance program. Founded in 2011, Kraken is a centralized global virtual currency exchange where users can buy, sell, or hold cryptocurrencies. OFAC alleged that Kraken violated the Iranian Transactions and Sanctions Regulations by not preventing sanctioned jurisdictions from opening Kraken accounts. Specifically, OFAC alleges that "between approximately October 14, 2015, and June 29, 2019, Kraken processed 826 transactions, totaling approximately $1,680,577.10 on behalf of individuals who appeared to have been located in Iran at the time of the transactions." These individuals were able to open accounts because Kraken did not implement IP address blocking on its platform.

As part of the settlement, Kraken agreed to invest $100,000 in sanctions compliance controls. OFAC's settlement with Kraken perfectly illustrates the dangers of improper or incomplete compliance research. Companies must ensure that the wallets under their control comply with OFAC's requirements and accurately track the flow of cryptocurrencies through them.[126]

LESSONS LEARNED: If anyone worldwide can download your cryptocurrency wallets, your company must take at least these actions:

1. Set up a robust compliance program.
2. Ensure your wallets can't be downloaded from sanctioned jurisdictions.
3. Report any suspicious activity in your wallets by filing Suspicious Activity Reports (SARs).

The next section will cover the internal revenue service's regulation of cryptocurrencies.

CRYPTOCURRENCIES AND THE INTERNAL REVENUE SERVICE

I believe the Internal Revenue Service (IRS) is the most significant government agency for US citizens regarding cryptocurrencies. Other government agencies, such as the SEC, CFTC, DOJ, and Treasury, regulate and enforce against companies and people who violate regulations and

[126] US Department of Treasury. "Settlement Agreement between the US Department of the Treasury's Office of Foreign Asset Controls and Payward, Inc. (Kraken)." November 28, 2022. https://ofac.treasury.gov/recent-actions/20221128#:~:text=Kraken%20agreed%20to%20remit%20%24362%2C158.70,in%20certain%20sanctions%20compliance%20controls. (Accessed October 28, 2025)

commit crimes. In contrast, the IRS regulates US citizens who invest in or use cryptocurrencies. The IRS requires all cryptocurrency investors, miners, and owners to report their cryptocurrency holdings, especially profits and losses. Thus, IRS regulations and sanctions directly affect everyday cryptocurrency users, miners, and investors. As such, it is essential to stay updated on IRS cryptocurrency policies and notices.

SCENARIO: You bought cryptocurrencies in 2016. You haven't purchased any cryptocurrencies since then, nor have you touched your cryptocurrency profits. Since you haven't engaged in any cryptocurrency investments in the past year, do you need to report owning cryptocurrencies to the IRS?

IRS NOTICE 2014-12 CLARIFICATIONS

After publishing Notice 2014-12, the IRS issued several notices clarifying its position on cryptocurrencies, stablecoins, and NFTs.

Digital assets are broadly defined as any digital representation of value that is recorded on a cryptographically secured distributed ledger or any similar technology as specified by the Secretary.

Digital assets include (but are not limited to):

1. Convertible virtual currency and cryptocurrency.
2. Stablecoins.
3. Non-fungible tokens (NFTs).[127]

Taxable gain or loss can result from transactions, including, but not limited to:

1. Sale of a digital asset for fiat.
2. Exchange of a digital asset for property, goods, or services.

---

[127] Digital Assets. IRS. IRS.gov, September 16, 2025. https://www.irs.gov/businesses/small-businesses-self-employed/digital-assets (Accessed October 28, 2025)

3. Exchange or trade of one digital asset for another digital asset.
4. Receipt of a digital asset as payment for goods or services.
5. Receipt of a new digital asset as a result of a hard fork.
6. Receipt of a new digital asset as a result of mining or staking activities.
7. Receipt of a digital asset as a result of an airdrop.
8. Any other disposition of a financial interest in a digital asset.
9. Receipt or transfer of a digital asset for free (without providing any consideration) that does not qualify as a bona fide gift.
10. Transferring a digital asset as a bona fide gift if the donor exceeds the annual gift exclusion amount.

This list is instructive for cryptocurrency investors and users to determine their tax liability. The IRS analysis of NFTs will be covered in Chapter 7.

CRYPTOCURRENCIES AND THE 1040 FORM

In April 2024, the IRS published FS-2024-12 titled: "Taxpayers Must Report Crypto and Other Digital Asset Transactions on their Tax Return." FS-2024-12 sheet reminds taxpayers that they must report any income from digital asset transactions on their tax returns, including sales, exchanges, and receipts as payment for goods or services. Here are some key points from the FS2024-12:

- Taxpayers must answer the question about digital assets on their tax return.
- Taxpayers must report all digital asset-related income, including gains, losses, and ordinary income, on their tax returns.
- Taxpayers should keep records of their digital asset transactions to support their tax filings.

If you buy, sell, own, or mine cryptocurrencies, it is essential to thoroughly review all IRS policies and documentation to fully understand your tax responsibilities. Additionally, consulting with a CPA or a tax expert with experience in cryptocurrency can help you avoid penalties.

In 2021, the IRS included a question about virtual currency at the top of Forms 1040, 1040-SR, and 1040-NR. All of these forms ask: "At any time during 2021, did you receive, sell, exchange, or otherwise dispose of any financial interest in any virtual currency?" The IRS also provides the following instructions:

> All taxpayers filing Form 1040, Form 1040-SR, or Form 1040-NR must check one box answering "Yes" or "No" to the virtual currency question. The question must be answered by all taxpayers, not just taxpayers who engaged in a transaction involving virtual currency in 2021.[128]

According to the IRS, taxpayers can check "no" under the following conditions:

1. Holding virtual currency in their own wallet or account.
2. Transferring virtual currency between their own wallets or accounts.
3. Purchasing virtual currency using real currency, including purchases using real currency on electronic platforms such as PayPal and Venmo.
4. Engaging in a combination of holding, transferring, or purchasing virtual currency as described above.

Taxpayers must check "yes" under the following conditions:

1. The receipt of virtual currency as payment for goods or services provided.
2. The receipt or transfer of virtual currency for free (without providing any consideration) does not qualify as a bona fide gift.
3. The receipt of new virtual currency from mining and staking activities.

---

[128] Internal Revenue Service. "Updates to Questions on Digital Assets; Taxpayers Should Continue to Report All Digital Asset Income." Internal Revenue Service, January 24, 2023. https://www.irs.gov/newsroom/irs-updates-to-question-on-digital-assets-taxpayers-should-continue-to-report-all-digital-asset-income. (Accessed September 3, 2025)

4. The receipt of virtual currency as a result of a hard fork.
5. An exchange of virtual currency for property, goods, or services.
6. An exchange/trade of virtual currency for another virtual currency.
7. A sale of virtual currency.
8. Any other disposition of a financial interest in virtual currency.

Remember that failing to report your cryptocurrency holdings or transactions is a direct violation of tax laws. This can lead to significant penalties, including fines and potential imprisonment.[129]

Finally, the IRS provides the following instructions:

- If a taxpayer disposed of any virtual currency held as a capital asset through a sale, exchange, or transfer, they must check "Yes" and use Form 8949 to calculate their capital gain or loss and report it on Schedule D (Form 1040).
- If a taxpayer received any virtual currency as compensation for services or disposed of any virtual currency that they held for sale to customers in a trade or business, they must report the income as they would report other income of the same type (for example, W-2 wages on Form 1040, 1040-SR, or 1040-NR, line 1, or inventory or services from Schedule C on Schedule 1).[130]

LESSON LEARNED 1: Any sale, exchange, or disposal of cryptocurrency, including for goods or services, is a taxable event. In other words, when you

---

129 McClure, Zac. What Happens if I Don't File My Crypto Taxes. Token Tax. TokenTax.co, May 2, 2025. https://tokentax.co/blog/dont-file-crypto-taxes# (Accessed October 28, 2025)

130 Internal Revenue Service. "IRS Reminds Taxpayers They Must Check a Box on Form 1040, 1040-SR or 1040-NR on Virtual Currency Transactions for 2021." (Accessed March 18, 2021). https://www.irs.gov/newsroom/irs-reminds-taxpayers-they-must-check-a-box-on-form-1040-1040-sr-or-1040-nr-on-virtual-currency-transactions-for-2021. (Assessed September 3, 2025)

sell, trade, or dispose of cryptocurrency, you must report any resulting capital gain or loss.

LESSON LEARNED 2: The IRS requires taxpayers to use Form 8949 to report the sale or exchange of digital assets. This form requires you to provide details such as the date acquired, date sold, basis, proceeds, and gain or loss. Individual taxpayers must pay capital gains taxes on the sale of cryptocurrency transactions. For example, taxpayers who receive free cryptocurrencies have to check "yes" on Form 1040, Form 1040-SR, or Form 1040-NR and report the free cryptocurrency.

LESSON LEARNED 3: If you invest, own, mine, or receive free cryptocurrencies, you should consult with a CPA or tax advisor with cryptocurrency experience to assess your tax liability.

INTERNAL REVENUE SERVICE COURT CASES

SCENARIO: Your cryptocurrency staking rewards are deposited into your wallet. You do not spend them, convert them into fiat currency, or deposit them into your bank, PayPal, or another type of account. Do you still have to pay taxes on your staking rewards?

JOSHUA JARRET V. THE UNITED STATES

At the time of publication, only one IRS court case addresses the taxation of cryptocurrency staking rewards. *In Joshua Jarrett v. United States*, Jarrett requested a refund with interest after the IRS taxed him for income received from staking his Tezos coins. Jarrett argued that rewards in the form of newly minted coins should only be taxable when they are sold, not when they are created. The IRS issued the refund, but Jarrett declined it and sued the IRS.[131]

The United States District Court for the Middle District of Tennessee dismissed Jarrett's case against the Internal Revenue Service (IRS), holding that the matter was rendered moot following the issuance of the refund to Jarrett by the IRS.

A US District Court decision would have presented a legal opinion and greater clarity on this stakeholder issue. For now, it appears the IRS practice is not to tax cryptocurrencies that have been staked unless they are sold.

LESSON LEARNED: As of now, staked cryptocurrency that is not spent, converted to fiat currency, or deposited into a bank account is not taxable. You are strongly advised to consult with a CPA or tax professional with cryptocurrency experience with cryptocurrencies to determine your tax liability.

SUMMARY

Regulating the cryptocurrency industry requires a delicate balance between fostering innovation and protecting investors. This chapter provides a comprehensive overview and updated information on regulation and enforcement, enabling everyone involved with cryptocurrencies — businesses, investors, miners, users, legislators, and, of course, attorneys — to gain a greater understanding of cryptocurrency regulations.

Here are some recommendations on how you can use the information in this chapter:

[131] Joshua Jarrett v. United States, No. 22-6023, 6th Cir. 2023.

- Invest in cryptocurrencies that are registered with the relevant government agencies that define their status and standards for compliance.
- Use cryptocurrency exchanges that prioritize customer service and investor protection. This could include requirements for transparent disclosure, safeguards against fraud and scams, and mechanisms for dispute resolution.
- Conduct research and utilize cryptocurrency exchanges that are registered and compliant with AML and KYC procedures. If possible, prioritize using cryptocurrency exchanges in your country before using an offshore exchange.

The regulation of cryptocurrencies has a significant impact on individual investors, shaping their opportunities, risks, and experiences within the cryptocurrency market. Investor protection is a common and important theme, ensuring that individuals are shielded from abusive practices, scams, and Ponzi schemes. Measures such as Know Your Customer (KYC) and Anti-Money Laundering (AML) regulations protect investors and deter illicit activities within the cryptocurrency ecosystem.

The above information and examples further indicate one common theme — that cryptocurrency regulations and sanctions are constantly evolving. A case in point is the evolution of cryptocurrency regulation from the Biden Administration to the Trump Administration, with no one knowing how the next administration will regulate cryptocurrencies. Investors must be proactive in reviewing regulatory requirements and adapt their investment strategies accordingly. Engaging with reputable platforms and exchanges. Conduct thorough due diligence and seek professional advice to mitigate risks and enhance investment outcomes. Ultimately, striking a balance between regulatory compliance, investor protection, and encouraging innovation will create a better and safer cryptocurrency market.

SUCCESS STRATEGY ACTION ITEMS

1. Thoroughly research what cryptocurrency transactions are legal or illegal in your country before you invest in a cryptocurrency. The results of the research may change your investment approach. On the global level, Comply Advantage lists updated legislation for numerous countries.

2. Read the Digital Assets Executive Order, the Digital Assets Fact Sheet, The 2025 Digital Asset Executive Order and Fact Sheet, and the Bitcoin Reserve Executive Order. These documents will provide insight into the regulatory and enforcement visions of two presidential administrations.
3. If you live or conduct business in the US, stay updated about the legislation that is introduced and passed at the national and state levels. Use Congress.gov (https://www.congress.gov) for national legislation and The National Conference of State Legislatures for state legislation.

4. Another option is to use online legal databases to search for any specific laws or regulations that address cryptocurrencies. For example, Westlaw, LexisNexis, or government websites that provide access to legislative texts.
5. Research any cryptocurrency you are considering for investment in the US on SEC.gov. Even if you do not find any information or enforcement actions on your cryptocurrency, SEC.gov will give you a lot of valuable

information. Reading SEC regulatory enforcement actions will teach you what kind of cryptocurrency companies and investments you should avoid. This research may help you avoid losing money on a bad cryptocurrency investment.

6. Visit the SEC Crypto Assets and Cyber Enforcement Website. When you visit this website, you will learn about details on cases brought against individuals and companies for fraudulent or unregistered crypto asset offerings and cyber misconduct. Information and resources to help identify and avoid crypto scams and understand related risks.

7. The CFTC, similar to the SEC, is actively enforcing against cryptocurrency companies that fail to register with the SEC or make false and misleading claims. If you are considering investing in a cryptocurrency business, especially one that promotes forex or commodities investments, research the company on the CFTC enforcement website. Nfafutures.gov is an excellent resource for researching the legitimacy of a cryptocurrency. Use the NFA Basic website https://www.nfa.futures.org/basicnet/ and the CFTC website cftc.gov/check. www.consumerfinance.gov or calling (855) 411-CFPB (2372).

8. Read all IRS notices and tax forms related to cryptocurrencies, including: IRS Notice 2014-12, 2014-12 Clarifications, and all IRS forms with cryptocurrency information. If you do not file your taxes correctly, the IRS may audit your tax filing. You are also strongly encouraged to consult a tax professional with cryptocurrency experience for tax advice.

CHAPTER 5

CRYPTOCURRENCIES AROUND THE WORLD

"Fiat currencies divide the world because they are based on policies, politics, and greed. Decentralized cryptocurrencies financially unite the world because they are based on mathematics and technology, empowering people financially."

Eric Guthrie, Esquire

Author, *Blockchain or Die (2019) and Blockchain or Die Harder (2026)*

As the son of a Jamaican father and a Bajan mother, I grew up in a vibrant, multicultural household with a unique tapestry of perspectives and traditions. Remittances were more than just money to us — they were a symbol of my parents' love, support, and emotional ties to their home countries. I remember my father sending money orders to Jamaica in those distinctive red, white, and blue airmail envelopes. Every single time he sent one, he'd rue the high transaction fees and lament how his money was being sent to corporations instead of our family. Years later, when cryptocurrencies offered legitimate alternatives to expensive remittance systems, I was fascinated by how this new system of sending international value would unfold. Imagine my delight when I saw that Jamaica is at the forefront of the global cryptocurrency scene, as you will learn in this chapter.

CHAPTER OVERVIEW

The theme of this chapter summarizes something I have been preaching for over seven years and truly believe in — the global unifying power of decentralized cryptocurrencies. There are very few fiat currencies that can be used in more than one country. Cryptocurrencies present viable alternatives to traditional global financial systems. Unless you are using a major credit card, you have to visit an ATM, a currency exchange, or a hotel front desk to acquire local currencies. In my international travels, I've seen "We accept Bitcoin" or "We Accept Cryptocurrencies" signs all over the world, and expensive exchange rates do not apply to cryptocurrencies.

Cryptocurrencies have become a trillion-dollar asset class. This explosive growth has required governments to explore ways to regulate the sale and use of cryptocurrencies in their countries. As more countries adopt cryptocurrencies, the market will become stronger and more valuable, potentially leading to increased cryptocurrency prices and greater opportunities for investors.

However, governments worldwide are still grappling with how to classify and regulate cryptocurrencies, which creates uncertainty and presents legal challenges for businesses and cryptocurrency investors.

Some governments are going beyond cryptocurrency regulation; they are creating central bank digital currencies (CBDCs), adopting existing cryptocurrencies, and establishing cryptocurrency exchanges. According to the International Monetary Fund (IMF), as of October 2025, Bermuda, Jamaica, and Nigeria have already implemented a CBDC, and an estimated 100 global governments are exploring CBDCs.

This chapter will cover global governments that have actually passed cryptocurrency legislation or created a CBDC, in chronological order.[132]

---

[132] Adrian, Tobias; He, Dong; Mancini-Griffoli, Tommaso; Sun, Tao. Central Bank Digital Currency Development Enters the Next Phase. IMF Blog, November 20, 2023. https://www.imf.org/en/Blogs/Articles/2023/11/20/central-bank-digital-currency-development-enters-the-next-phase#:~:text=The%20Bahamas%2C%20Jamaica%2C%20and%20Nigeria,Kingdom%20are%20at%20the%20forefront. (Accessed September 3, 2025)

Many governments are still considering or debating the passage of legislation or the creation of a CBDC. These government actions will not be covered in this chapter.

The next section will cover the jurisdictions that passed cryptocurrency legislation: Bermuda, the Bahamas, El Salvador, the European Union, and Georgia

CRYPTOCURRENCY LEGISLATION

2018-BERMUDA

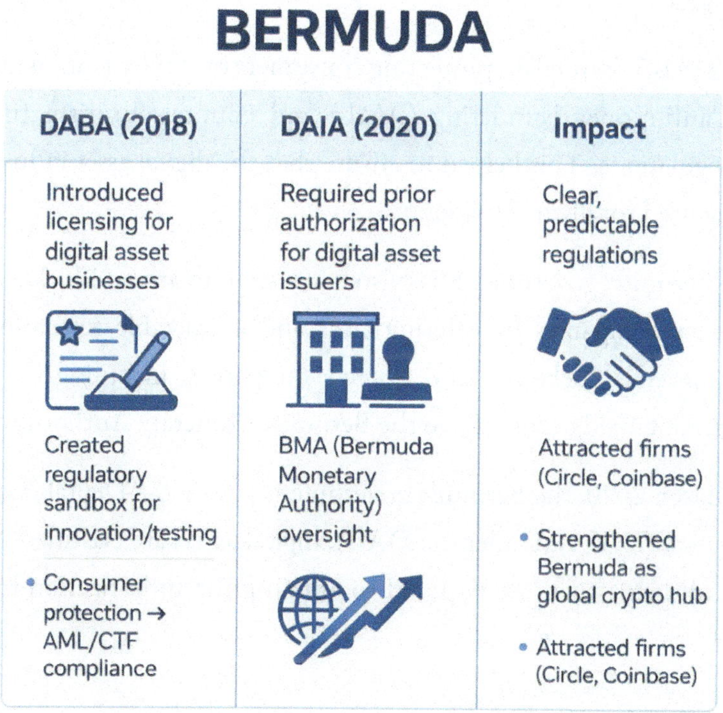

Bermuda was one of the first countries to officially embrace cryptocurrencies by passing legislation, regulations, and business initiatives

to attract cryptocurrency companies. In 2018, the Bermuda legislature passed the Digital Asset Business Act (DABA) to regulate and provide a legal framework for digital asset businesses.[133]

The DABA introduced a licensing regime for businesses operating in the digital asset space. This provided a level of oversight and regulation, ensuring that companies complied with specific standards. The DABA also included a regulatory sandbox—a controlled environment where businesses can test innovative products or services under limited regulatory oversight, fostering innovation and allowing regulators to observe new technologies in action. This enabled experimentation with new digital asset technologies and business models, helping to mitigate risks associated with emerging technologies.

The DABA focused on protecting consumers by implementing measures such as anti-money laundering (AML) and counter-terrorism financing (CTF) regulations. This helped to ensure that the digital asset industry was not being used for illicit activities.

The Bermuda government has also taken steps to attract blockchain and digital asset companies by offering clear and predictable regulations and policies. Several blockchain and fintech companies, including Circle and Coinbase, obtained licenses from the Bermuda Monetary Authority.

In March 2020, the Bermuda government passed the Digital Assets and Issuance Act (DAIA). Under the DAIA, digital assets are classified as BMA assets, and their issuers are required to obtain prior authorization from the BMA.[134]

---

[133] Digital Asset Business Act 2018: 28

[134] Digital Assets and Issuance Act 2020: 18

I see the passing of the DABA and DAIA as truly positive milestones for the growth of blockchain technology and Bermuda's economy. These laws establish a clear regulatory framework for digital asset issuances, helping reduce uncertainty and encouraging businesses and investors to participate. They also promote innovation and economic development by creating a welcoming environment for the digital asset industry and inviting new companies to Bermuda. Ultimately, these laws strengthen Bermuda's position as a global leader in digital asset regulation.

Of course, passing these laws is only the beginning of the story. The real test of Bermuda's success in digital regulation will hinge on how companies and industries respond to the legislation and Bermuda's enforcement actions. I know several attorneys and policymakers who are closely monitoring the implementation of both acts to analyze the potential challenges that may arise. Nevertheless, Bermuda's regulatory and enforcement actions will have a great influence over the global cryptocurrency space.

2020-BAHAMAS

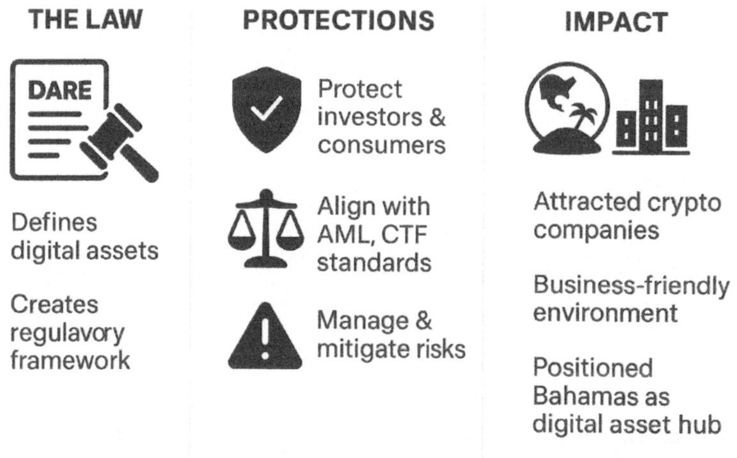

In December 2020, the Bahamas legislature passed the Digital Assets and Registered Exchanges (DARE) Act, which created a legal definition for digital assets and a regulatory framework for digital asset businesses and activities permitted in the Bahamas.

The DARE Act established a regulatory framework that was designed to:

1. Protect investors and consumers.
2. Align with international standards to combat money laundering, the financing of terrorism, and weapons proliferation; and manage and mitigate attendant risks.[135]

The DARE Act established a regulatory framework to facilitate digital asset innovation and foster its development in the Bahamas.

The passing of the DARE Act was one of the reasons numerous companies considered the Bahamas for their corporate headquarters. For example, Sam Bankman-Fried initially set up FTX in Hong Kong, but China passed legislation and made statements that did not benefit the cryptocurrency market. Fried moved FTX to the Bahamas due to the stable legislation and regulatory framework, as well as the absence of corporate and income taxes, which created a welcoming business environment.[136]

---

[135] The Security Commission of the Bahamas. Consultation on the Digital Assets and Registered Exchanges Bill, 2023. The Securities Commission of the Bahamas. Scb.gov.bs, April 25, 2023. https://www.scb.gov.bs/wp-content/uploads/2023/04/DARE-Bill-2023-Consultation-Document.pdf (Accessed October 28, 2025)

[136] Ibid.

2021-EL SALVADOR

On September 7, 2021, the Salvadoran government passed "Ley Bitcoin," which translates to "Bitcoin Law" in English. The Salvadoran government passed Ley Bitcoin for several reasons, including economic stability, financial inclusion, and remittances.[137] Ley Bitcoin made Bitcoin the official national currency for El Salvador alongside the US dollar. El Salvador's reliance on the US dollar made it vulnerable to fluctuations in the value of the US dollar. Ley Bitcoin provides a decentralized alternative that could boost El Salvador's economy. The law also waives capital gains for the buying and selling of Bitcoin. [138]

Ley Bitcoin stirred global interest, raising questions about El Salvador's use of Bitcoin as a fiat currency, the future of cryptocurrencies, and their potential impact on traditional financial systems.[139]

---

[137] Jose' Ignacio Hernandez. In El Salvador, Bitcoin's Retreat Left Valuable Lessons. Americas Quarterly. AmericaQuarterly.org, March 17, 2025. https://www.americasquarterly.org/article/in-el-salvador-bitcoins-retreat-left-valuable-lessons/#:~:text=In%20September%202021%2C%20El%20Salvador,those%20tied%20to%20new%20technologies. (Accessed October 28, 2025)

[138] Tuttle, Brad. "What Making Bitcoin an Official National Currency Means for El Salvador." Money.com, January 14, 2022. https://money.com/bitcoin-el-salvador-national-currency/ (Accessed September 30, 2025)

[139] El Salvador Congress Approves Bitcoin as Legal Tender. Aljazeera. Aljazeera.com, June 9, 2021. https://www.aljazeera.com/news/2021/6/9/el-salvador-congress-approves-bitcoin-as-legal-tender (Accessed October 28, 2025)

There are major financial benefits to El Salvador's adoption of Bitcoin. Approximately 70 percent of their population is unbanked or underbanked, making financial accessibility a pressing issue. With the nationwide adoption of Bitcoin, all a citizen needs is a cell phone and internet access to trade Bitcoin, unlocking financial freedom and limitless opportunities. By potentially increasing financial inclusion, reducing remittance fees, and promoting greater economic stability, the adoption of Bitcoin could foster significant economic growth in El Salvador.[140]

Despite these financial benefits, El Salvador's adoption of Bitcoin as a fiat currency has faced significant challenges. The implementation of the state-backed Bitcoin wallet, Chivo, encountered technical issues during its launch, leading to public dissatisfaction and doubts about adopting Bitcoin as a currency. In addition, many Salvadorans expressed concerns about Bitcoin's volatility, fearing that fluctuations in its value could negatively impact their savings and El Salvador's economy.[141]

In December 2024, El Salvador agreed to "confine" Bitcoin use as part of a $1.4 billion loan agreement with the IMF. This involved restrictions on cryptocurrency transactions and the use of state-backed wallets. El Salvador continues to hold Bitcoin in its reserves, demonstrating its continued commitment to the cryptocurrency.[142]

---

[140] Arsianian, Henri; Donovan, Robert; Blumenfeld, Matthew; Zamore, Anthony. El Salvador's Law: A Meaningful Test for Bitcoin. PWC.com. https://www.pwc.com/gx/en/financial-services/pdf/el-salvadors-law-a-meaningful-test-for-bitcoin.pdf (Accessed October 28, 2025)

[141] El Salvador Ends Bitcoin as a Legal Tender – Here's Why Bitcoin Didn't Work. Newsroompanama.com, February 7, 2025. https://newsroompanama.com/2025/02/07/el-salvador-ends-bitcoin-as-legal-tender-heres-why-bitcoin-didnt-work/ (Accessed October 29, 2025)

[142] da Silva, João. World's First Bitcoin Nation Scales Back Crypto Dream. BBC. BBC.com, December 14, 2024. https://www.bbc.com/news/articles/c20e3l3xllwo (Accessed October 28, 2025)

2023-EUROPEAN UNION: MICA

Europe distinguishes itself with proactive regulation. The Markets in Crypto-Assets (MiCA) framework (2023) provides a unified set of EU rules for tokens, exchanges, and stablecoins. Germany leads in institutional adoption, authorizing crypto funds for professional investors, while France pushes for integration with traditional banking. Europe's strength lies in its clarity, which reduces uncertainty for startups and investors.

MiCA also includes market integrity provisions to prevent market abuse and insider dealing. MiCA is the first multi-government legislation of its stature, and other national legislative bodies will likely learn from the steps taken to pass it, as well as from how the EU enforces it.

The following framework illustrates how the EU's MiCA regulation translates legal clarity into market impact.

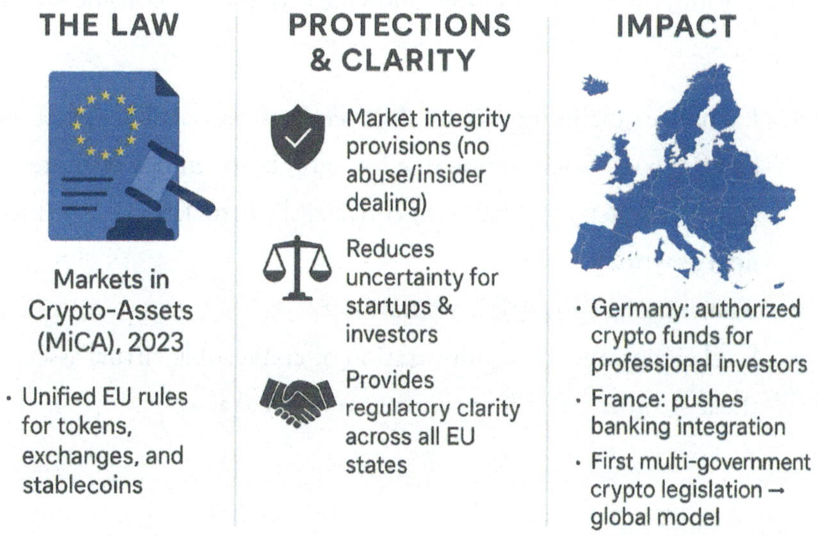

2023-GEORGIA: VASP

In June 2023, the National Bank of Georgia (NBG) issued a regulation for Virtual Asset Service Providers (VASPs).[143] The NBG provides the following definitions for virtual assets and convertible virtual assets:

- "A virtual asset is a digital representation of value that is interchangeable and non-unique. It can be digitally traded or transferred and used for investment purposes and/or payment. Virtual assets do not include digital representations of fiat currencies, securities, or other financial instruments."[144]

- "A convertible virtual asset is a virtual asset that has an equivalent value on the market, in national or foreign currency, in another virtual asset, and/or a financial instrument in which it can be exchanged.[145]"

In addition, the NBG states the following activities are considered virtual asset services:

- Exchange (including via kiosks) between convertible virtual assets and fiat currencies (national or foreign), between one or more forms of virtual assets, and between convertible virtual assets and financial instruments.
- Transfer of a convertible virtual asset.
- Safekeeping and/or administration of convertible virtual assets or of the instrument enabling control over virtual assets.

---

[143] National Bank of Georgia. "Virtual Asset Service Providers (VASPS)." https://nbg.gov.ge/en/page/virtual-asset-service-providers-vasps (Accessed September 3, 2025)
[144] Ibid.

[145] Ibid.

- Portfolio management of convertible virtual assets (excluding collective portfolio management).
- Administration of the trading platform of the convertible virtual assets.
- Lending of convertible virtual assets.
- Initial Coin Offering of convertible virtual asset and/or service related to the initial coin offering.[146]

The following figure summarizes Georgia's 2023 VASP regulatory framework, outlining the scope of services covered and its broader regulatory impact.

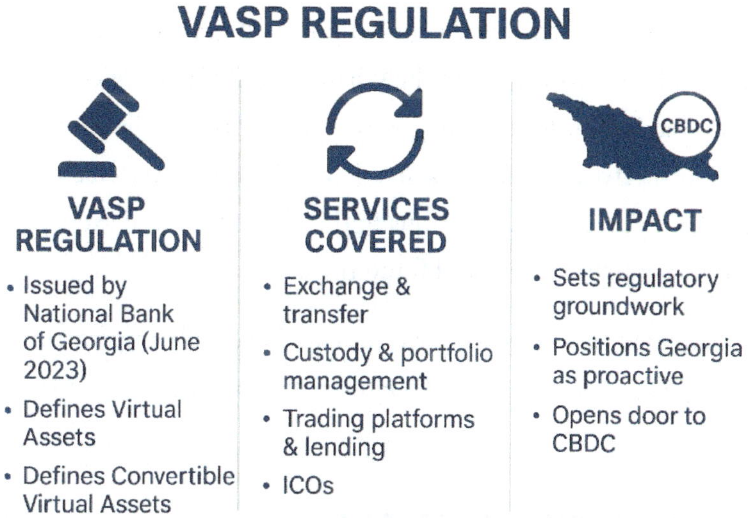

The above legislation set the groundwork for other countries to pass and implement their own legislation. However, legislation is only the beginning of the journey; once it is passed, country leaders must build the infrastructure

[146] Ibid.

to support the legal use of cryptocurrencies. One possible path is for the country to implement a CBDC.

CENTRAL BANK DIGITAL CURRENCIES

As covered in Chapter 1, Central Bank Digital Currencies (CBDCs) are a revolutionary concept that has the potential to reshape the global financial landscape. CBDCs promise to provide a more efficient, secure, and inclusive payment system, while also addressing challenges such as financial inclusion and the implementation of monetary policy. As central banks around the world explore the development and deployment of CBDCs, it is crucial to understand their potential benefits, risks, and broader implications for global financial systems. Africa is the global leader in grassroots crypto adoption. Nigeria ranks #2 worldwide in the Chainalysis 2023 adoption index, despite a central bank ban. Kenya integrates crypto with mobile money (M-Pesa), while South Africa enforces exchange regulation.[147] Yet challenges remain, such as high currency fluctuations in the global financial exchange markets, weak internet penetration, and fraud risks.

As covered below, three governments have fully implemented Central Bank Digital Currencies (CBDCs): the Bahamas, Nigeria, and Jamaica.

2020-BAHAMAS: SAND DOLLAR (CBDC)

Sand Dollar is the digital version of the Bahamian dollar (B$), which is issued by the Central Bank of The Bahamas through authorized financial institutions (AFIs). While the Sand Dollar is issued by the Central Bank of

[147] The Chainanalysis Team. The 2023 Global Crypto Adoption Index: Central and Southern Asia Are Leading the Way in Grassroots Crypto Adoption. Chainanalysis.com, September 12, 2023. https://www.chainalysis.com/blog/2023-global-crypto-adoption-index/ (Accessed November 30, 2025)

the Bahamas, the Bahamian Government also collaborated with the International Monetary Fund (IMF) to ensure more thorough oversight of the process.[148]

The following figure highlights the Bahamas' 2020 Sand Dollar CBDC, summarizing its issuance model, key features, and intended impact.

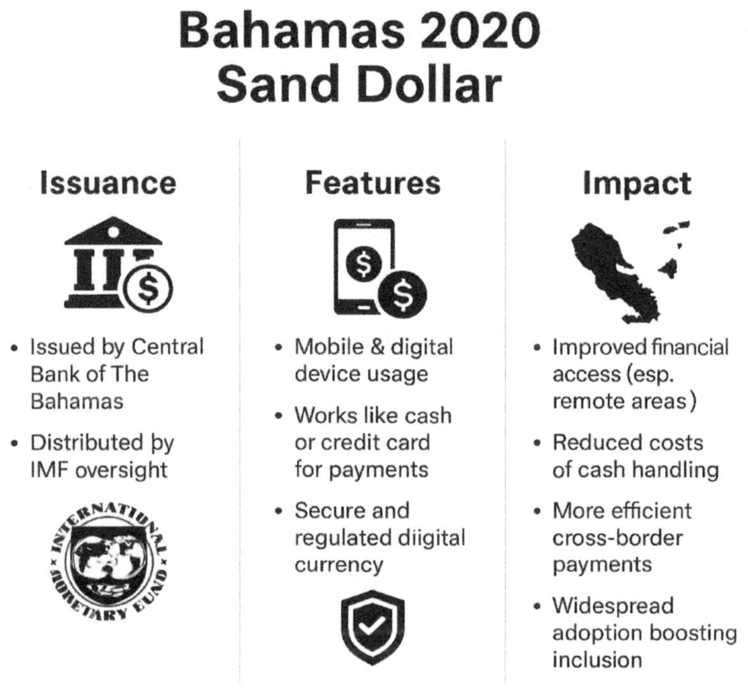

The Sand Dollar is used through mobile phones and other digital devices. People can use it to make payments, just like they would with cash or a credit card. So, what are the advantages of a Sand Dollar over Bahamian dollars? Here are a few salient points:

[148] The Bahamas: The World's First Digital Currency. IMF.org, March 1, 2021. https://www.imf.org/en/Videos/view?vid=6238535912001 (Accessed October 30, 2025)

- Improved access to financial services: People in remote areas can access financial services more easily.
- Reduced costs: Using the Sand Dollar can reduce the costs associated with using cash, such as transportation and storage.
- Increased efficiency: The Sand Dollar can improve the efficiency of cross-border payments.[149]

The Sand Dollar has been widely adopted by businesses and consumers in the Bahamas, helping to improve financial inclusion in the country. In a YouTube Video, Shaqueno Porter, Deputy Manager, Financial Market Infrastructure Unit, Currency Department of the Central Bank of the Bahamas, provided an update on the progress of the Sand Dollar, including the following data.[150]

The following graph shows the Sand Dollars in circulation over the past three years:

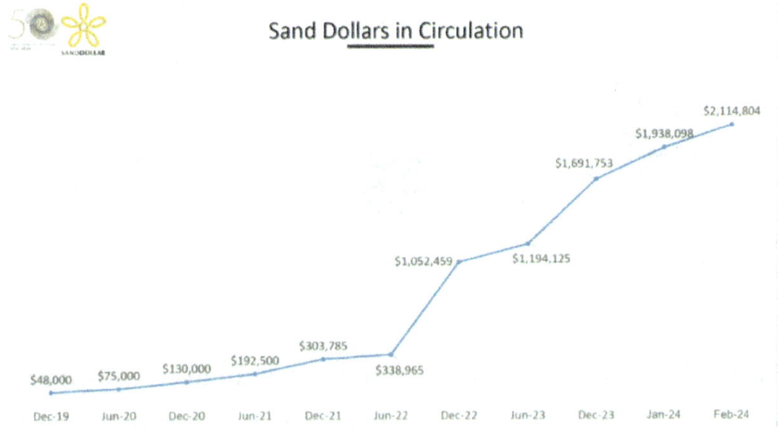

---

[149] About Us. Sanddollar.bs. https://www.sanddollar.bs/about#:~:text=Sand%20Dollar%20is%20a%20central,backed%20by%20the%20foreign%20reserves. (Accessed October 28, 2025)

[150] An Update on the Sand Dollar by Shaqueno Porter. Youtube.com, December 2024. https://www.youtube.com/watch?v=cS57rju0SUI (Accessed October 30, 2025)

Increase in Sand Dollar Wallets Over the Past Three Years

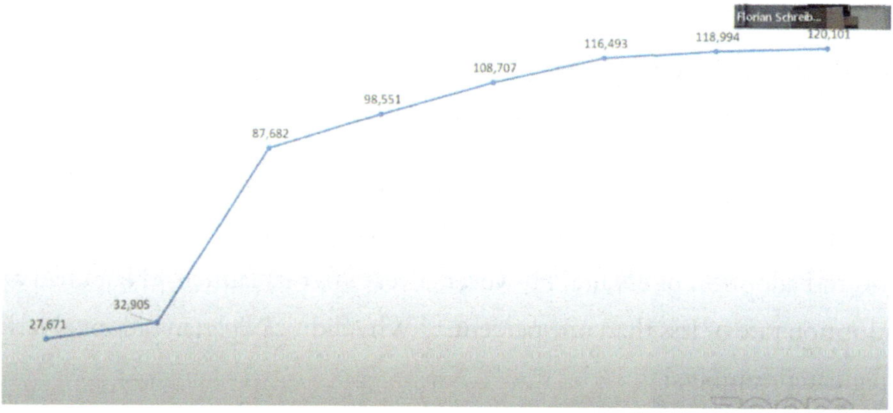

The Bahamian government reports that the country comprises 700 islands, of which only 30 are inhabited. The Sand Dollar has been particularly beneficial for residents of remote islands in the Bahamas, providing easier access to financial services and reducing reliance on physical cash. By offering digital payment options, the Sand Dollar could increase financial inclusion for the unbanked and underbanked.

The Bahamas is prone to natural disasters. In fact, when Hurricane Dorian made landfall in 2019, I was checking in on friends and associates in the Bahamas, and I remember a few of them mentioned the Sand Dollar. At the time, I did not realize the Sand Dollar was a CBDC. Nevertheless, the Sand Dollar, which the economy introduced when access to traditional banks or financial institutions became nearly impossible, was introduced. The IMF reports that during this time, the Sand Dollar helped many residents and communities, enabling essential services to continue operating, especially on remote islands.

2021-NIGERIA: THE ENAIRA CBDC

In October 2021, the Central Bank of Nigeria (NCB) released eNaira, the Nigerian CBDC, which was designed to:

1. Modernize Nigeria's payment system.
2. Improve financial inclusion.
3. Reduce transaction costs.
4. Enhance the efficiency of monetary policy.[151]

According to Chainalysis, Nigeria has the highest cryptocurrency adoption rate in Africa.[152] Based on this, analysts predicted the widespread use and adoption of eNaira. However, a year after its launch, eNaira had an adoption rate of less than one percent.[153] Why didn't Nigerians adopt eNaria in greater numbers?

Unlike the decentralized Bitcoin and Ethereum blockchains, eNaira is a CBDC that is strictly controlled by the NCB. The avoidance of centralized control is a defining characteristic of cryptocurrencies and the feature that initially attracts most users to them. Therefore, a government-controlled centralized CBDC may not be their digital currency of choice.

The following figure outlines Nigeria's 2021 launch of the eNaira CBDC, highlighting its rollout, core features, and key implementation challenges.

---

151 Preface. Why the Bank of Nigeria Made the Cash-less Policy. Central Bank of Nigeria. CBN.gov.ng, https://www.cbn.gov.ng/cashless/Cash-Less%20FAQs.pdf (Accessed October 30, 2025)

152 Chainanalysis. "Cryptocurrency Penetrates Key Markets in Sub-Saharan Africa as an Inflation Migration and Trading Vehicle." September 19, 2023. https://www.chainalysis.com/blog/africa-cryptocurrency-adoption/ (Accessed September 3, 2025)

153 Salami, Iwa, "eNaira: Nigeria's Digital Currency has had a Slow Start – What's Holding in Back." SundayMail.co, February 20, 2024, https://www.sundaymail.co.zw/enaira-nigerias-digital-currency-has-had-a-slow-start-whats-holding-it-back (Accessed September 3, 2025)

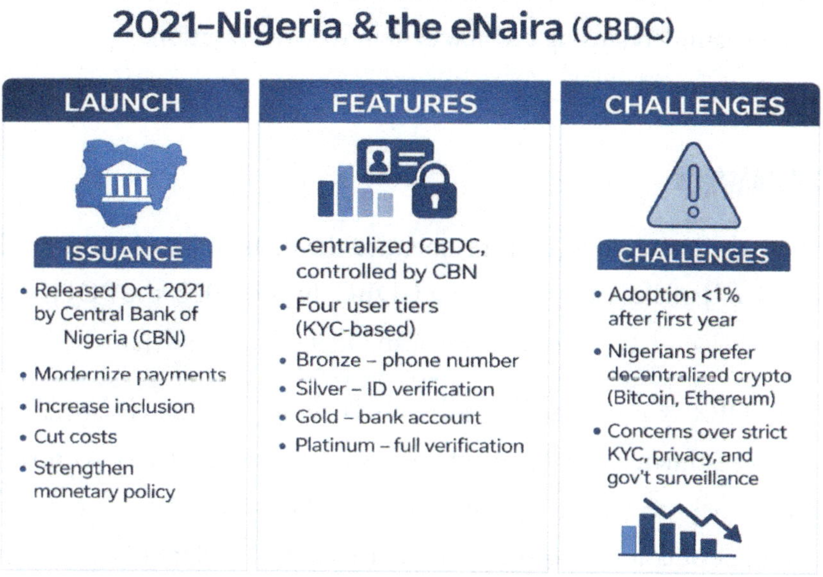

Another reason for the relative failure of eNaira could arguably be its strict requirements. Here's a breakdown of the user tiers likely based on KYC (Know Your Customer) requirements:

- Tier 0 (Bronze): $120k (approx. $80 USD), requires basic verification (phone number).
- Tier 1 (Silver): $300k (approx. $200 USD), requires additional ID verification.
- Tier 2 (Gold): $500k (approx. $330 USD), requires bank account details.
- Tier 3 (Platinum): $5m (approx. $3300 USD), strictest verification, including utility bills.[154]

While higher tiers offer greater access to capital, they also require users to provide additional information. Users may consider eNaira's strict KYC/AML requirements a breach of privacy, with the government able to monitor all transactions. This, along with the nature of cryptocurrency

---

[154] Your eNarina Wallet is in Your Control, eNarina. eNarina.gov.ng. https://enaira.gov.ng/for-individuals/ (Accessed October 28, 2025)

ledgers, makes every single transaction traceable and ensures that money cannot be laundered using the new system to avoid taxation.[155]

2022-JAMAICA

In 2022, the Jamaican legislature amended the Bank of Jamaica Act to allow the Bank of Jamaica (BOJ) to issue a CDBC. In July 2022, the BOJ issued its non-blockchain-based CBDC, Jamaican Digital Exchange (Jam-DEX).

Jam-DEX is positioned to solve challenging Jamaican financial access issues. It is estimated that 17 percent of the Jamaican population is unbanked, and a higher percentage is underbanked. However, 55 percent of the Jamaican population reportedly has internet access, and mobile phone usage is 100 percent. The high cell phone and internet use gives Jamaicans the financial incentive and the technological ability to use Jam-DEX.[156] Therefore, Jam-DEX can have a profound impact on the Jamaican population.

However, Jam-DEX has not been widely adopted by the Jamaican population. None of the Jamaicans I spoke with, whether in the US or Jamaica, had heard of Jam-DEX. Even during a recent trip to Jamaica, not a single person—restaurant staff or store owners alike—was familiar with it.

The following figure summarizes Jamaica's 2022 Jam-DEX CBDC, highlighting its issuance model, core features, and early adoption challenges.

---

[155] Lawal, Temitayo. Nigeria's Digital Currency Can't Compete with Crypto. Rest of World. RestofWorld.org, May 17, 2023. https://restofworld.org/2023/nigeria-digital-currency-failing/ (Accessed October 28, 2025)

[156] Gail, Elizabeth. "Jamaica's Central Bank Digital Currency and the Problems it Hopes to Solve." Cointelegraph.com, March 25, 2022, https://cointelegraph.com/news/jamaica-s-central-bank-digital-currency-and-the-problems-it-hopes-to-solve (Accessed August 30, 2025)

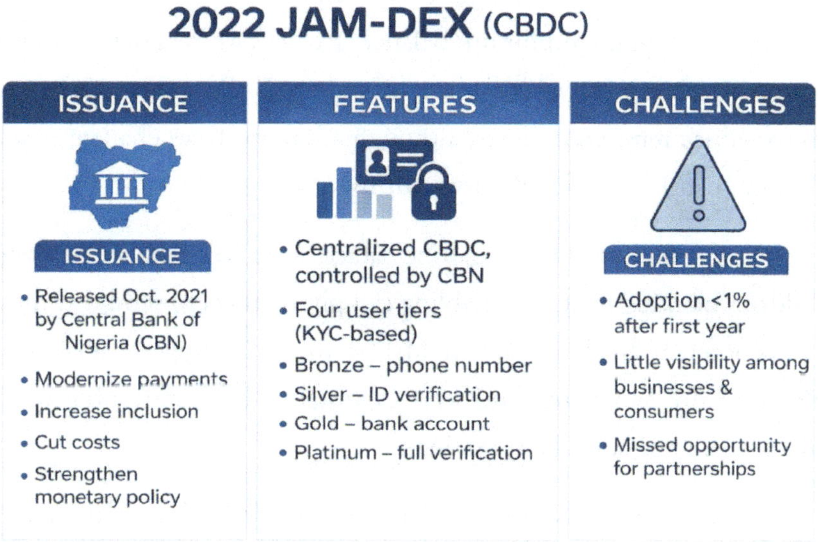

THREE COUNTRY CBDC ANALYSIS

After thorough research and analysis of each of these three CBDCs, through their websites, social media, and press coverage, I reached the conclusion that the initial user adoption of the CBDCs in each country would have been more impactful if the respective central banks had partnered with major companies or industries in their country and persuaded employers to use the CBDCs to pay their employees' wages. Paying wages in a CBDC would, in turn, require employees to download a wallet, receive their wages in the CBDC, and then use it for rent or mortgage payments, bills, and everyday purchases. Alternatively, the employee could convert the CBDC to cash, but that would still constitute a CBDC transaction. This would significantly increase the nationwide adoption of the respective CBDCs.

There is a history of widespread use of a digital currency in a country. The story of M-Pesa is worth exploring in this regard. As I discussed in *Blockchain or Die*, M-Pesa is a digital currency that started in Kenya in 2008. Many rural families lack access to banks because they do not meet the Kenyan banking regulations' identification requirements or do not have sufficient funds to maintain the minimum balance. Kenya's M-Pesa demonstrates the success of mobile money; blockchain pilots (e.g., BitPesa, now AZA Finance) aim to reduce remittance costs, although adoption faces challenges such as currency volatility and regulatory scrutiny.

M-Pesa was created by Vodafone Safaricom, a Kenyan mobile operator, to address banking issues by enabling cell phone users to send text messages to each other. Individuals can use M-Pesa to receive wages, transfer money internationally, buy goods and services, obtain a loan, start a business, and make any other financial transactions.

According to CNN.com's "M-Pesa: Kenya's Mobile Money Success Story Turns 10," in 2016, 30 million users across 10 countries processed 6 billion transactions at a peak rate of 529 per second. This is truly a user adoption success story.[157]

M-Pesa's implementation as a digital currency was ahead of its time. Since M-Pesa is not used as an investment, it is more useful as a means of payment, which benefits the Kenyan people and economy. In another conversation with a friend who owns a Kenyan restaurant outside of Nairobi, I asked if she accepted M-Pesa. Her response was, "M-Pesa is life. My business could not survive if I didn't use M-Pesa."

The success story of M-Pesa is an example for governments on how to design CBDC user adoption models to increase CBDC wallet downloads and usage.

---

[157] Monks, Kieron. M-Pesa: Kenya's Mobile Money Success Story Turns 10. CNN.com, February 17, 2024. https://edition.cnn.com/2017/02/21/africa/mpesa-10th-anniversary (Accessed January 24, 2026)

THE FUTURE OF CBDCS

This section of the chapter covers fully implemented CBDCs. As of February 2025, the Atlantic Council reports that 134 countries and currency unions are in the process of implementing their own CBDCs, whether in the development or pilot phases. The Atlantic Council reports the following CBDC development numbers:

- Pilot phase – 49
- Development phase – 19
- Research phase – 36

That is a total of 102 CBDCs across the three phases. These numbers show the impact of digital currencies on the global economy and promise a future of limitless potential for the global cryptocurrency industry.[158]

WHERE ARE CRYPTOCURRENCIES BANNED?

This chapter would be incomplete without discussing countries where governments have passed laws or regulations banning cryptocurrencies.

General Cryptocurrency Ban (Absolute Ban):

A general ban means that all cryptocurrency activities are prohibited within a country's jurisdiction. This typically includes:

- Trading: Buying, selling, or exchanging cryptocurrencies.
- Mining: The process of creating new cryptocurrency units.
- Holding/Possession: Owning or storing cryptocurrencies.
- Usage as Payment: Using cryptocurrencies for goods or services.
- Cryptocurrency Exchanges: Running platforms for crypto trading.

---

[158] Atlantic Council Geoeconomics Center. Central Bank Digital Currency Tracker. The Atlantic Council, August 14, 2025. https://www.atlanticcouncil.org/cbdctracker/ (Accessed September 3, 2025)

Countries with general bans are usually attempting to completely eradicate cryptocurrency activity, often with severe penalties. Some countries even impose criminal penalties for non-compliance. According to the Atlantic Council, at the time of publication, the following countries have a general, or absolute, ban.

- Algeria
- Bangladesh
- China
- Egypt
- Ethiopia
- Morocco
- Nepal
- Pakistan
- Saudi Arabia
- Tunisia[159]

In countries with absolute bans, possessing, trading, or mining cryptocurrencies can lead to severe legal penalties, including fines and imprisonment. Investors and users must be aware of these risks to prevent accidental violations. One such example is China.

China's stance is paradoxical: it banned crypto trading and mining in 2021, pushing operations abroad, yet simultaneously promotes the Blockchain Service Network (BSN) and pilots the Digital Yuan (e-CNY) across major cities. This reflects Beijing's priority on state control over decentralized systems, positioning blockchain as an arm of governance rather than a means of financial freedom.

---

[159] Atlantic Council. Cryptocurrency Regulation Tracker. The Atlantic Council. https://www.atlanticcouncil.org/programs/geoeconomics-center/cryptoregulationtracker/ (Accessed September 3, 2025).

CRYPTOCURRENCY AND BLOCKCHAIN DEGREES

The final section of this chapter offers a more futuristic perspective on global cryptocurrencies, specifically exploring the potential role that colleges and universities can play in their future.

Institutions of higher learning, of which many are government run or subsidized, shape the leaders of tomorrow, and college curricula play a decisive role in that process. These future leaders will include politicians and policymakers, technology and business executives, venture capitalists, media professionals, law enforcement officials, and the next generation of college and university faculty. As more colleges and universities introduce blockchain-focused courses and degree programs in cryptocurrency and blockchain technology, these emerging leaders will develop the knowledge and practical skills needed to design and implement impactful solutions for businesses, governments, and the global community.

At the time of publication, only a handful of colleges offer courses in cryptocurrency and blockchain technology, and even fewer offer degrees in these fields. The following colleges and universities offer cryptocurrency and blockchain technology degrees:

Associates Degree

- Atlantic International University – Associate in Blockchain Technology and Digital Currency
- Union College of Union County, New Jersey – Blockchain Technology A.A.S.

Bachelors

- Morgan State University – Blockchain Technology Major and Minor
- Albany State University – Blockchain Technology Minor

Masters

- University of the Cumberlands – Online Master's in Global Business with Blockchain Technology
- Northeastern University – Master of Science in Information Systems - Blockchain Technology Specialization
- Fordham University – MBA with Blockchain Secondary Concentration
- Atlantic International University - Master's in Blockchain Technology and Digital Currency
- University of Nicosia – Blockchain and Digital Currency, MSc

Doctorate

- Bircham International University – PhD in Cryptocurrency
- Erasmus Research Institute of Management – PhD in Business and Management – Cryptocurrency and Blockchain Specialization
- University of Camerino – International School of Advanced Studies – PhD Blockchain and Distributed Ledger Technology

This is only the beginning. The list of colleges and universities offering degrees in cryptocurrency and blockchain is likely to expand. Many schools offer courses or certificate programs, so if you're short on time for a full degree, taking individual classes or certificates could be a great choice.

SUMMARY

This chapter covered the burgeoning era of global cryptocurrency legislation and CBDCs. As the world transitions towards a more digital economy, CBDCs may emerge as a solution to bridge the gap between traditional financial systems and the innovative realm of cryptocurrencies.

We discussed the potential benefits of CBDCs, such as reduced transaction costs, increased financial stability, and improved traceability. Ultimately, this chapter paints a picture of a future where CBDCs could revolutionize how we interact with money. As governments and central banks continue to explore this groundbreaking technology, the global adoption of CBDCs is likely to revolutionize the financial landscape in the decades to come.

Finally, the chapter highlights the significance of colleges and universities incorporating blockchain technology and cryptocurrencies into their curricula and degree programs. Individuals who graduate with these degrees and complete these courses will become future leaders in cryptocurrency and blockchain. Blockchain technology will likely play a meaningful role in parts of the global economy, but the pace and scope of its adoption remain uncertain. The coming decade will reveal whether it becomes foundational infrastructure or remains a specialized tool.

SUCCESS STRATEGY ACTION ITEMS

1. Follow the Central Bank in countries of interest. The Atlantic Council CBDC Tracker is an excellent, user-friendly tool for staying up to date on global CBDC adoption.

2. Follow global cryptocurrency legislation. Countries that pass progressive cryptocurrency legislation are more likely to create a conducive environment for cryptocurrency and blockchain companies. The Atlantic Council is an excellent and user-friendly tool to stay up to date on global cryptocurrency legislation.

Follow the adoption of cryptocurrencies in colleges and universities. Once again, when colleges and universities begin teaching cryptocurrency and blockchain courses and eventually offer degree programs, the country will have the talent to create more cryptocurrency and blockchain businesses and attract more tech investors.

Part II
PRESENT APPLICATIONS AND FUTURE POSSIBILITIES

CHAPTER 6

BLOCKCHAIN TECHNOLOGY UPDATES

"Make no mistake, blockchain will be a game changer, but it is not the answer to every business opportunity or problem." [160]

Julie Sweet

Chairwoman and Chief Executive Officer of Accenture.

CHAPTER OVERVIEW

The growth of cryptocurrencies is closely linked to the development of blockchain technology. Bitcoin, the first and most well-known cryptocurrency, was built on the foundation of blockchain technology—a decentralized and transparent ledger system.

As Bitcoin and cryptocurrencies grew in popularity, interest in the underlying blockchain technology also increased. While the public focused on cryptocurrencies, business leaders and investors recognized blockchain's broader potential as a transformative technology with applications across sectors such as supply chain management, healthcare, and voting systems. This awareness has led to a surge in blockchain projects and applications designed to address industry-specific challenges.

---

[160] Brainy Quote. BrainyQuote.com. https://www.brainyquote.com/quotes/julie_sweet_1104454#:~:text=Julie%20Sweet%20Quotes&text=Make%20no%20mistake%2C%20blockchain%20will%20be%20a%20game%20changer%2C%20but,every%20business%20opportunity%20or%20problem. (Accessed September 4, 2025)

Julie Sweet's insightful quote resonates deeply with the current state of blockchain technology: while it holds immense potential, many companies are attempting to create blockchain business solutions for non-existent business challenges. Another factor to consider is whether blockchain technology is currently mature enough to solve specific business challenges. For blockchain to reach its full potential, adoption rates and user interfaces for blockchain applications must improve significantly.

Overall, the decentralized finance (DeFi) sector continues to expand, with new protocols and services emerging. Yield farming, lending and borrowing platforms, decentralized exchanges (DEXs), and synthetic asset platforms are gaining popularity. The total invested in DeFi protocols has reached billions of dollars, indicating increased user adoption and interest.

Blockchain business solutions have also undergone significant evolution since the publication of *Blockchain or Die* in 2019. This chapter will delve deeper into the major developments of numerous blockchain companies, focusing on the underlying technology and innovative ways it is being used to create new business solutions. This chapter will also explore important advancements, including Level 2, Level 3, and cross-chain blockchains. Additionally, it will examine how corporations and governments have adopted blockchain technology.

INTRODUCTION

When I began offering training programs on cryptocurrencies and blockchain technology, I quickly realized the growing interest from businesses and organizations eager to understand this revolutionary field. However, during these sessions, I faced a curious challenge: my primary client contacts often asked me to omit Bitcoin from the curriculum. They cited concerns about its perceived risks and associations with illegal activities, fearing it might influence their attendees' investment choices. This intriguing hesitation highlights the complex perceptions surrounding Bitcoin and blockchain today. I always responded with two succinct points: first, historical analysis does not constitute investment advice. Second, while it is true that Bitcoin and other cryptocurrencies have been, and continue to be, used to commit crimes, the same can be said for fiat currencies. The US Dollar, the Euro, and other currencies have historically been used to facilitate the commission of crimes totaling trillions of dollars. After this conversation, my primary client contacts always realize that discussing Bitcoin during training was necessary for attendees to completely understand blockchain history and functionality.

The history of Bitcoin is also crucial to understanding blockchain technology, as it introduced the proof-of-work (PoW) protocol. PoW solved the "double-spending" problem, ensuring that digital assets could not be duplicated and spent only once. PoW required miners to expend computational power to verify transactions and create new blocks, making it difficult for malicious actors to manipulate the network.

However, the PoW protocol has slow processing speeds and consumes significant energy; therefore, blockchain companies that followed Bitcoin have developed newer, faster, and more energy-efficient protocols. It is from here that we start our journey on blockchain technology updates.

A SUMMARY OF BLOCKCHAIN PROTOCOLS

Blockchain protocols are the foundational rules and procedures that govern how a blockchain network operates — the blueprint that defines how information is recorded, verified, and shared across the blockchain network.

If Bitcoin and Ethereum solved old puzzles, today's researchers are tackling the unsolved ones: privacy, governance, and scale. Zero-Knowledge Proofs (zk-SNARKs, zk-STARKs) enable someone to prove a fact without revealing the underlying data. On Zcash, this means you can prove you own the money without showing your balance. On zkSync or StarkNet, it means compressing thousands of transactions into a single proof, boosting speed without sacrificing security. The promise is huge, but so are the trade-offs: high computational costs and limited auditability.

Meanwhile, Decentralized Autonomous Organizations (DAOs) are testing with new models of collective governance. MakerDAO manages billions of dollars in stablecoin reserves through token votes. At the same time, ConstitutionDAO famously raised $47 million in an attempt to buy a copy of the U.S. Constitution—but still fell short.[161]

The ideal is transparency; the reality is that whale investors and voter apathy often skew outcomes.

Modular blockchains. Instead of doing everything on one chain (Ethereum's model), modular systems like Celestia and Fuel split execution, consensus, and data availability into separate layers. Think of it as moving from a single all-purpose computer to a network of specialized servers. This makes scaling easier, but it creates new challenges around coordination and security.

And the frontier keeps pushing outward: cross-chain bridges like Wormhole, privacy coins like Monero, and experimental consensus models like Proof-of-History, such as those used by Solana or Avalanche. Each test strikes a balance between innovation and risk. The field is no longer a monolith; it's a laboratory with breakthroughs and failures happening side by side.

---

[161] Sigalos, MacKenzie. The Crypto Investors Who Raised $47 Million for But a Copy of Constitution Lost Their Bid – Here's Where the Money Goes Now. CNBC.com, November 19, 2021. https://www.cnbc.com/2021/11/18/constitutiondao-crypto-investors-lose-bid-to-buy constitution-copy.html#:~:text=A%20consortium%20of%20crypto%20investors,to%20purpose%20the%20remaining%20funds. (Accessed October 28, 2025)

The following are the most common blockchain protocols.

PROOF OF WORK (POW) PROTOCOL: BITCOIN, DOGECOIN, AND LITECOIN

Historically, PoW is the most common type of blockchain protocol. It involves miners competing to solve a complex mathematical problem. The winner gets to add a new block to the blockchain and receives a reward.

PROOF OF STAKE (POS) PROTOCOL: ETHEREUM, CARDANO, AVALANCHE

Unlike PoW, the PoS protocol does not require miners to solve complex mathematical problems. Instead, the protocol selects the next validator to add a new block to the blockchain based on how much cryptocurrency they hold and are willing to "stake." The miner's stake in the cryptocurrency increases their odds of adding a new block to the blockchain and receiving the award. The more cryptocurrency the miner owes, the more likely they are to be selected to add a new block to the blockchain.

PoS has been widely accepted as a safer, lower-energy protocol. It is also less prone to hacking because there is less incentive to hack a platform where users are staking their cryptocurrencies.

DELEGATED PROOF OF STAKE (DPOS) PROTOCOL: EOS, BITSHARE, STEEMIT

This protocol is similar to PoS, but instead of assigning the right to add a new block based on the amount of cryptocurrency a miner holds, it assigns the right based on the number of votes the miner receives from other network participants. The more votes a miner has, the more likely they are to be selected to add a new block to the blockchain.

PROOF OF AUTHORITY (POA): VECHAIN AND POA NETWORK

PoA delegates the authority to validate transactions to a predefined set of nodes, unlike Proof of Work (PoW) or Proof of Stake (PoS), which rely on computational power or token holdings. This results in faster processing, but it also means the system is more centralized, as only trusted validators are allowed to produce blocks.

BYZANTINE FAULT TOLERANCE (BFT) PROTOCOL: HYPERLEDGER FABRIC

This protocol allows blockchains to continue functioning even in the presence of malicious or malfunctioning nodes. BFT ensures that all honest nodes agree on the order of transactions and blocks added to the blockchain, thus maintaining network security and reliability even in cases of partial failures.

DIRECTED ACYCLIC GRAPH (DAG) PROTOCOL: IOTA, NANO, HEDRA HASHGRAPH

This protocol enables high transaction throughput and scalability. Instead of using blocks, DAG-based blockchains use a graph structure to store and confirm transactions. Each transaction confirms previous ones, resulting in high throughput and scalability—ideal for projects that demand a large transaction volume and real-time processing.

HYBRID PROTOCOL: RIPPLE

Some blockchains use a hybrid protocol, which combines features of multiple protocols. For example, the Ripple blockchain uses a hybrid protocol that combines elements of PoW and BFT protocols. By acting as a "bridge currency," XRP facilitates the rapid exchange of fiat currencies, reducing transaction times and costs. This has the potential to significantly enhance the speed and efficiency of international trade and remittances, thereby making financial services more accessible to a broader population.

Now that we've covered these seven blockchain protocols, let's take a deeper dive into specific blockchain updates.

THE "PROOF OF STAKE" ETHEREUM

Since 2020, many new blockchain companies have chosen to use protocols other than PoW. However, Ethereum's shift from PoW to PoS, often called "The Merge," was a significant development, supporting the application and use of the PoS protocol.[162]

Ethereum's founder, Vitalik Buterin, initially launched Ethereum using PoW, but after seven years, he recognized that PoW was less energy-efficient and more susceptible to attacks than PoS. Starting in 2020, Ethereum updated its protocol, switching from PoW to PoS to create Ethereum 2.0 (Eth 2.0). Eth 2.0 was a major change for Ethereum and Ethereum miners, as Eth 2.0 transactions are faster, more secure, and more efficient.[163] Eth 2.0 enhances scalability, security, and sustainability.

ETHEREUM 2.0 AND GODADDY

In February 2024, GoDaddy partnered with Ethereum Name Service (ENS) to connect domain names and cryptocurrency wallets. According to GoDaddy, ENS "allows crypto wallets, apps, and services to link to a domain name via DNS without having to pay gas fees and requiring no additional costs, enabling domains to act as web addresses for the wallets themselves." GoDaddy customers can connect their domain name with their ENS-compatible crypto wallet, including Coinbase Wallet, MetaMask, Rainbow, Uniswap Wallet, and more. [164]

---

[162] Fidelity Viewpoints. What is the Ethereum Merge? Fidelity. Fidelity.com, May 7, 2025. https://www.fidelity.com/learning-center/trading-investing/ethereum-merge (Accessed October 28, 2025)

[163] Ethereum.com. https://ethereum.org/en/roadmap/ (Accessed September 30, 2025)

[164] "GoDaddy and Ethereum Name Service Bridge the Gap Between Domain Names and Crypto Wallets." GoDaddy.net, February 05, 2024, https://aboutus.godaddy.net/newsroom/company-news/news-details/2024/GoDaddy-and-Ethereum-Name-Service-Bridge-the-Gap-Between-Domain-Names-and-Crypto-Wallets/default.aspx (Accessed September 3, 2025)

This collaboration merges GoDaddy's dominance in web domains with Ethereum's innovation in decentralized apps (DApps), DAOs, and NFTs—paving the way for user-friendly blockchain integration for businesses and individuals. This partnership combines established technology with new advancements to create innovative products for companies and individuals.

CARDANO HARD FORK

In August 2021, Cardano announced the launch of the Alonzo hard fork, which led to a 116% surge in Cardano's price the following month. On September 12, 2021, the Cardano Alonzo hard fork was officially launched, bringing smart contract functionality to the Cardano blockchain. The Alonzo hard fork enables developers to build decentralized applications (dApps) on the Cardano blockchain. Over 100 smart contracts were deployed within 24 hours of the launch. Cardano's adoption of smart contracts provides blockchain businesses with an alternative to the Ethereum blockchain.[165]

SOLANA UPGRADE

In early 2022, Solana released the Solana 1.9 upgrade, which included improvements to its blockchain's stability, security, and scalability, reportedly making it faster and more efficient.[166]

In April 2024, Solana 1.17.31 was another significant update aimed at alleviating network congestion issues.

---

[165] McCall, Isaiah. "Cardano's Alonzo Hard Fork: Everything you Need to Know," Medium.com, August 19, 2021, https://medium.com/yardcouch-com/cardanos-alonzo-hard-fork-everything-you-need-to-know-a73aa0082af3 (Accessed October 28, 2025)

[166] Willing, Nicole. "Binance Smart Chain (BSC)." Technopedia, June 9, 2023. https://www.techopedia.com/definition/binance-smart chain-bsc#:~:text=Binance%20Smart%20Chain%20implements%20the,applications%20(dApps)%20like%20MetaMask. (Accessed September 3, 2025)

In May 2025, Solana launched the Alpenglow upgrade. Alpenglow is a major overhaul of Solana's consensus protocol, designed to address bottlenecks related to finality and voting overhead.

Although Ethereum is the second-largest blockchain by market capitalization, it is also the most expensive in terms of gas/transaction fees. I have noticed that Solana and Cardano have gained popularity among many blockchain businesses due to their lower transaction fees. Lower fees translate to a lower bar for a project to reach profitability. They may also make it more affordable for blockchain businesses to give away products, like NFTs, to build brand awareness and gain market share.

BINANCE SMART CHAIN UPGRADE

Binance's primary blockchain ecosystem, known as BNB Chain, has undergone several significant upgrades recently, primarily focusing on enhancing performance, scalability, and preparing for future innovations, especially in AI and advanced dApp functionalities. Here's a summary of the key recent upgrades:

BNB CHAIN FUSION (CONSOLIDATION OF BNB BEACON CHAIN)

The older BNB Beacon Chain (BEP-2 standard), which handled staking and governance, has been fully migrated and merged into the BNB Smart Chain (BSC, BEP-20 standard).[167]

Chain Fusion is designed to simplify the BNB Chain architecture, enhance efficiency, reduce security risks by operating on a single, more robust chain, and prepare for future technological demands. For users, this

---

[167] BNB Chain Beacon Chain Migration Guidance. BMBChain.org, May 13, 2024. https://www.bnbchain.org/en/blog/the-bnb-chain-beacon-chain-migration-guidance (Accessed September 3, 2025)

means that all BNB Chain activities, including holding BNB, staking, and interacting with dApps, now occur on the BNB Smart Chain (BEP-20).

PASCAL HARD FORK (MARCH 2025)

The Pascal hard fork streamlines the user experience, enables more complex and efficient dApp interactions, and paves the way for future advancements, such as gasless transactions across multiple tokens. The hard fork enables features such as batch approvals, making interactions smoother for users and providing more flexibility for developers building dApps. [168]

LORENTZ HARD FORK (APRIL 2025)

The Lorentz hard fork is designed to improve performance, reduce block times, and prepare the network for high-frequency, AI-powered decentralized services. The upgrade reduced BSC block times from 3 seconds to 1.5 seconds, and opBNB block times to 0.5 seconds. [169]

MAXWELL HARD FORK (MAY/JUNE 2025)

The Maxwell hard fork significantly reduces BSC's block times from 1.5 seconds down to 0.75 seconds. This is a crucial step towards "sub-second block times." This hard fork accelerates transaction confirmations, improves dApp responsiveness, and enables smoother experiences for DeFi and GameFi. [170]

---

[168] Turner, Liam. Pascal Hard Fork: BNB Chain's Next Big Step Towards Smart Contract Wallets & Ethereum Compatibility. CryptoNinjas.net. February 22, 2025. https://www.cryptoninjas.net/news/pascal-hard-fork-bnb-chains-next-big-step-towards-smart-contract-wallets-ethereum-compatibility/ (Accessed September 3, 2025)

[169] Lorentz Hard Fork – BSC FAQS. BNBChain.org. https://docs.bnbchain.org/bnb-smart-chain/faq/lorentz-hard-fork-upgrade/ (Accessed September 3, 2025)

[170] BNB Chain Announces Maxwell Hardfork: BSC Moves to 0.75-Second Block Times. BNBChain.org, May 22, 2025. https://www.bnbchain.org/en/blog/bnb-chain-announces-maxwell-hardfork-bsc-moves-to-0-75-second-block-times (Accessed September 3, 2025)

LAYER-2 BLOCKCHAIN SOLUTIONS

Although blockchain technology is revolutionary, decentralized blockchains face significant challenges—especially around speed and scalability. For example, the original Bitcoin blockchain requires about 10 minutes to verify a transaction. Since the launch of Bitcoin, developers have continually sought ways to enhance blockchain speed and scalability.

Think of Layer-1 blockchains as busy highways. They're great for long-distance travel, but can become congested during peak hours. Layer-2 solutions are like side streets or HOV lanes. They offer alternative routes that can alleviate traffic and reduce congestion on the main highway. This means faster transaction speeds and lower fees.

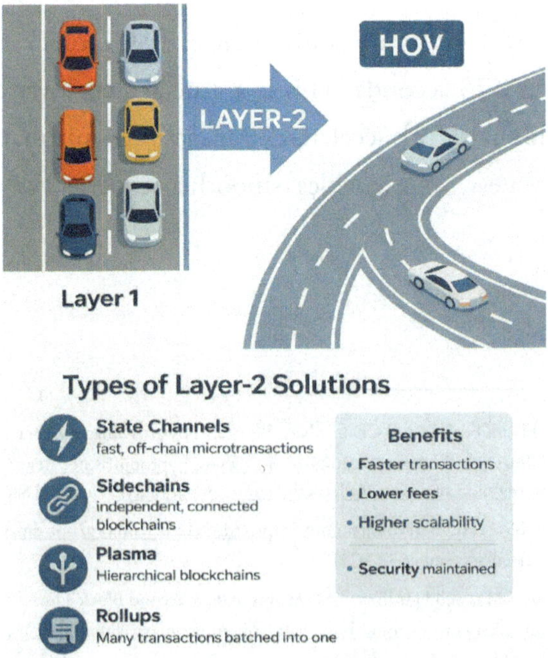

The above image shows how Layer-2s process transactions off Layer-1, then settle back to it—boosting throughput and lowering fees while preserving Layer-1 security.

The slow and congested PoW protocol required Bitcoin and other blockchains to find solutions to stay relevant and competitive with newer blockchain protocols. Consequently, numerous Layer 2 blockchain solutions have been created. Layer-2 blockchain solutions (Layer-2 solutions) are protocols or technologies built on top of existing blockchain networks, often called "Layer-1" blockchains, to enhance scalability, improve transaction speed, reduce costs, and address other limitations of the underlying blockchain.[171]

Layer-2 solutions enhance speed and scalability by offloading some transaction processing or computation to the off-chain while maintaining the security and trust of the Layer-1 blockchain. They enable higher transaction throughput and provide a better user experience without sacrificing the security guarantees of the underlying blockchain.

There are several types of Layer-2 solutions:

1. **State channels:** These are off-chain protocols that enable users to conduct transactions without relying on the main blockchain. Only the opening and closing transactions are recorded on the Layer-1 blockchain, while all the intermediate transactions are handled off-chain. State channels enable fast, low-cost transactions, making them ideal for applications such as microtransactions, gaming, and instant payments.
2. **Sidechains:** These are separate blockchains that are interoperable with the main blockchain. They enable users to transfer assets from the main blockchain to the sidechain, where transactions can be processed more

---

[171] Hurme, Oskar. "A Starter Guide to Using Layer 2 Blockchains." Block Spaces, January 19, 2023. https://www.blockspaces.com/blog/layer-2-blockchains-guide. (Accessed November 1, 2025)

quickly and at lower fees. Once the transactions are completed on the sidechain, the final outcome is recorded on the main blockchain. Sidechains reduce the load on the main blockchain while maintaining a strong connection to it, enabling scalability.

3. **Plasma:** A framework of smart contracts that supports hierarchical blockchains, in which multiple child chains are connected to a parent chain. It uses the security and consensus of the parent chain, while the child chains can process transactions more quickly and efficiently. Only the essential information is submitted to the parent chain.
4. **Rollups:** These are Layer-2 constructions that bundle multiple transactions together and submit a summary or proof of those transactions to the main blockchain. Rollups enhance scalability by consolidating multiple transactions into a single batch, thereby reducing costs and increasing transaction throughput.[172]

These Layer-2 solutions improve blockchain efficiency and scalability while retaining the security and decentralization properties of the underlying Layer-1 blockchain. They enable a wider range of applications and use cases that require faster, more cost-effective transactions and better user experiences.

LAYER 3 BLOCKCHAIN SOLUTIONS

Now we've established Layer-1 solutions as a wide, busy highway, and Layer-2 solutions as the various toll roads and side streets that offer a less congested alternative route. Layer-3 solutions are the smaller byways and alleys in these side streets that lead to specific destinations. In other words, Layer-3 solutions operate on top of Layer-2 solutions and primarily focus on

---

[172] What are 'Rollups' in Crypto. Coinbase.com. https://www.coinbase.com/learn/wallet/what-are-rollups-in-crypto (Accessed October 30, 2025)

particular applications or functionalities. They are more conceptual and less widely implemented than Layer-2 solutions.

Layer-3 solutions allow for customization of functionalities to cater to the specific needs of Layer-1 or Layer-2 (e.g., faster gaming transactions, enhanced privacy for DeFi). They also enable communication and data exchange between Layer-3 applications built on Layer-2 solutions.[173]

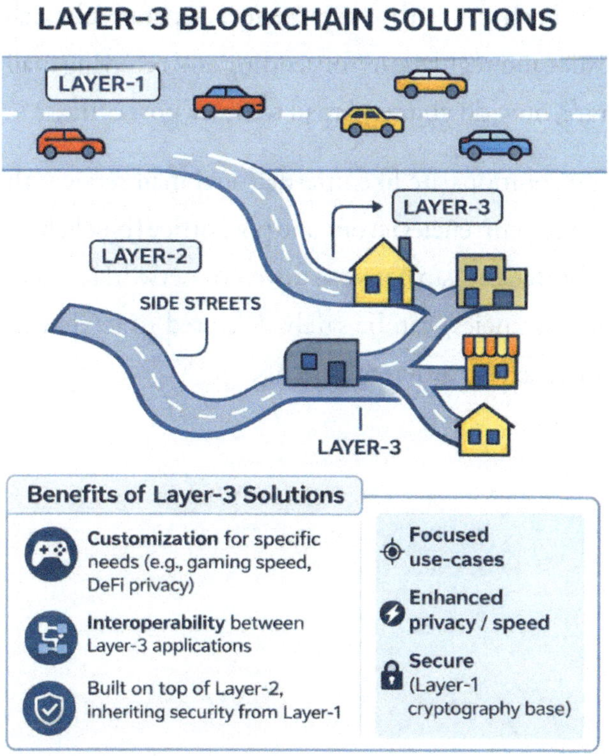

Layer-2 and Layer-3 blockchains remain secure as they both use the same cryptography as Layer-1 blockchains.

[173] Lepcha, Mensholong. "Layer 3 (L3) Blockchain." Technopedia.com, January 25, 2024. https://www.techopedia.com/definition/layer-3-blockchain (Accessed September 3, 2025)

CROSS-CHAIN SOLUTIONS

The prior section used the highway comparison to describe the functionality of Layer 1, 2, and 3 solutions; now, let's use the postal service to describe the efficiencies of cross-chain solutions.

Think of a blockchain solution as your local post office, and cryptocurrency as your mail. Now imagine a global network of post offices delivering mail. Each post office represents a separate blockchain network with its own rules and regulations for sending and receiving mail. If each post office is separate, how do you send mail to other post offices?

Cross-chain solutions are like international mail services that allow you to send mail (cryptocurrencies) from one post office (blockchain) to another. These services bridge the gap between different networks, ensuring that your mail (or cryptocurrencies) can be safely delivered to its destination, even if the two post offices have different systems.

CROSS-CHAIN INTEROPERABILITY

Different blockchain networks transferring assets and data between each other

Interoperability between different blockchain networks is a crucial development. Numerous blockchains, including Polkadot, Cosmos, and Avalanche, are developing cross-chain solutions that allow seamless communication and data transfer across multiple blockchain platforms, simplifying and expediting the exchange of assets and information.[174]

---

[174] Everything Blockchain. "Advancing Blockchain Interoperability." November 15, 2022. https://everythingblockchain.medium.com/comparing-advancing-blockchain-interoperability-polkadot-cosmos-and-avalanche-a42e548b5bf (Accessed September 30, 2025)

These updates highlight the ongoing advancements and real-world applications of blockchain technology. The field continues to evolve rapidly, with new developments and innovations emerging regularly.

DECENTRALIZED VS. CENTRALIZED BLOCKCHAIN ENTERPRISE SOLUTIONS

Now that we have covered the updates in protocols and hard forks, it is important to discuss the differences between decentralized and centralized blockchain solutions.

Decentralized and centralized blockchains differ fundamentally in their structure, control, and operational model. While both are ledgers that record information, the way they are managed and secured is entirely different.

Decentralized blockchains, such as Bitcoin and Ethereum, are distributed ledgers where control is not held by any single person or entity. It is managed by a network of participants (nodes) who collectively maintain the ledger.

- **Control:** Control is distributed among all participants in the network. There is no central point of authority.
- **Permissions:** Networks can be "permissionless," allowing anyone to join and participate in the network's consensus process.
- **Speed & Scalability:** These networks are generally slower and less scalable than centralized ones because they require a consensus mechanism among many nodes to validate transactions.
- **Transparency:** All transactions are publicly visible and immutable, providing a high degree of transparency.
- **Trust:** Trust is established through cryptographic proof and a consensus algorithm, not a central authority. This makes the network highly resistant to censorship and single points of failure.

Centralized Blockchains: A centralized blockchain is controlled by a single entity or a small group of entities. This control extends to validating transactions, adding new blocks, and enforcing the network's rules.

- **Control:** A single authority, like a corporation or government, has ultimate control over the network.
- **Permissions:** Access is typically "permissioned," meaning only authorized users can participate.
- **Speed & Scalability:** Centralized blockchains can be very fast and scalable because the network doesn't require a large number of nodes to reach a consensus.
- **Transparency:** The level of transparency is determined by the central authority. Data can be hidden or altered by the controlling entity.
- **Trust:** Trust is placed in the central authority. If the controlling entity is compromised or acts maliciously, the entire network is at risk.

In essence, the core difference lies in the distribution of power and speed. A centralized blockchain trades decentralization for efficiency, control, and speed, making it more suitable for private, enterprise-level applications. A decentralized blockchain, on the other hand, prioritizes security, transparency, and censorship resistance, making it the ideal model for public, trustless systems.

Unfortunately, many blockchain projects market their blockchains as "decentralized," but they are really centralized. I have been a consultant on many projects where the blockchain company claims to have developed a decentralized blockchain, but it's actually a centralized blockchain. The main difference lies in the "control."

When a company creates a blockchain solution and retains control of the protocols, the blockchain is centralized or, at the very least, a hybrid blockchain that combines decentralized and centralized components.

The next section will cover corporations adopting blockchain technology enterprise solutions, which are most certainly centralized blockchain enterprise solutions.

CORPORATIONS ADOPTING BLOCKCHAIN TECHNOLOGY

Many companies and corporations are actively exploring and implementing blockchain solutions. Numerous industries, including supply chain management, identity management, healthcare, finance, and logistics, are leveraging blockchain technology to improve transparency, traceability, and efficiency in their operations.

Several companies have used blockchain technology to improve efficiency, reduce costs, and enhance transparency. Some of these companies are household names, and you may be surprised to know that they have been committed to integrating blockchain technology into their business structure for many years. Listed in chronological order, a brief overview of corporations that implemented blockchain solutions follows:

2016-WALMART

Walmart partnered with IBM to implement blockchain technology solutions in its food supply chain management system. By leveraging blockchain to track the origin and movement of food products, Walmart has significantly improved food safety, cutting the time required to trace contaminated products from seven days to only 2.2 seconds.[175] The Walmart blockchain reduces food safety recalls and improves transparency for consumers.

I use the Walmart food supply management blockchain story as a case study in all my Blockchain Fundamentals Courses because it describes the

---

[175] Case Study. How Walmart Brought Unprecedented Transparency to the Food Supply Chain with Hyperledger Fabric. LDFDecentralizedTrust.org. https://www.lfdecentralizedtrust.org/case-studies/walmart-case-study (Accessed October 28, 2025)

power and benefits of blockchain technology in a context everyone understands: buying groceries. And not just any groceries; the desire to buy fresh, uncontaminated groceries. At the end of this story, attendees are usually left with a basic understanding of how blockchain technology is used in everyday life.

2016-VISA

Visa, a global payments technology company, has been actively exploring the potential of blockchain technology to revolutionize the payments industry. Here's a brief overview of their efforts:

Visa has also partnered with cryptocurrency platforms, including Coinbase and Binance, to enable payments in cryptocurrencies. This includes allowing businesses to accept cryptocurrencies as payment and converting them into fiat currency. These partnerships will make it easier for crypto users at all levels to buy, sell, and transact in cryptocurrencies.[176] I use Visa's blockchain adoption and partnership with cryptocurrency platforms as a case study in my blockchain training.

Finally, Visa has been actively developing secure blockchain solutions to meet the high standards of the payments industry. Of course, as cryptocurrency companies assert that they provide safer, more secure transactions, Visa must also provide a comparable or higher level of security for its users. Otherwise, no one will use the Visa platform.[177] Overall, Visa's adoption of blockchain technology is a strategic move to position itself as a leader in the future of payments.

---

[176] Helms, Kevin. Visa Partners with 60 Crypto Platforms to Let Customers Spend Digital Currency at 80 Million Merchants. Bitcoin.com. News.Bitcoin.com, December 21, 2021. https://news.bitcoin.com/visa-partners-with-60-crypto-platforms-let-consumers-spend-digital-currency-80-million-merchants/ (Accessed October 28, 2025)

[177] Ibid.

2016-JPMORGAN CHASE AND CONSENSYS

In 2016, JPMorgan Chase developed a Quorum blockchain platform for various purposes, including streamlining the settlement of securities transactions and issuing its own digital currency, JPM Coin. ConsenSys purchased Quorum from JPMorgan Chase.[178]

As the fifth-largest bank in the world by assets, JPMorgan Chase's entry into the blockchain space clearly indicates that large global financial institutions are paying close attention to blockchain technology. It can be predicted that major traditional companies and blockchain companies will soon start engaging in mergers and acquisitions. Eventually, these mergers and acquisitions may begin to consolidate the number of blockchain companies and actual blockchains to a more sustainable level.

2017-IBM

IBM has been a leader in blockchain technology, developing its own blockchain platform, IBM Blockchain. IBM has implemented blockchain solutions in various industries, including: supply also supply chain management, financial services, healthcare, and others.[179] In fact, according to the IBM Blockchain Newsletter, Home Depot, Renault, and Atea are using IBM blockchain.

---

[178] Allison, Ian. CoinDesk. "ConsenSys Acquires JPMorgan's Quorum Blockchain." Coin Desk, August 25, 2020. https://www.coindesk.com/business/2020/08/25/consensys-acquires-jpmorgans-quorum-blockchain/ (Accessed October 28, 2025)

[179] del Castillo, Michael. IBM Goes Live With First Commercial Blockchains. Coindesk.com, September 11, 2021. https://www.coindesk.com/markets/2017/03/20/ibm-goes-live-with-first-commercial-blockchains/ (Accessed October 29, 2025)

2018-AMAZON

In 2018, Amazon launched two blockchain-related services: Amazon Managed Blockchain and Amazon Quantum Ledger Database (QLDB).[180]

Amazon Managed Blockchain is a fully managed service that enables easy creation and management of scalable blockchain networks using open-source frameworks such as Hyperledger Fabric and Ethereum. It manages a decentralized network of nodes without worrying about the underlying blockchain infrastructure.[181]

Amazon QLDB is a fully managed ledger database that provides a transparent, immutable, and cryptographically verifiable transaction database. Amazon QLDB is designed to provide a transparent, tamper-proof method for storing data that can be audited in real time.[182]

In addition to these services, Amazon has also filed several blockchain-related patents, including one for a "proof-of-work" system that utilizes cryptography to secure transactions and prevent fraud, and another for a "Signaling System" that leverages blockchain technology to facilitate communication between autonomous vehicles.[183]

Amazon's entry into the blockchain space was a major development because, well, it's Amazon, one of the largest and most iconic global brands.

---

[180] Press Center. AWS Announces Amazon Quantum Ledger Database (QLDB) and Amazon Managed Blockchain. Amazon. Press.aboutamazon.com, November 28, 2018. https://press.aboutamazon.com/2018/11/aws-announces-amazon-quantum-ledger-database-qldb-and-amazon-managed-blockchain (Accessed October 28, 2025)

[181] Ibid.

[182] Amazon Patents System to Help Autonomous Vehicles Handle Traffic Shifts. Geekwire.com, January 17, 2017. https://www.geekwire.com/2017/amazon-patent-self-driving-reversible-lanes/#:~:text=The%20patent%2C%20issued%20today%2C%20concentrates%20on%20how,contact%20with%20a%20central%20roadway%20management%20system (Accessed October 28, 2025)

[183] Boyle, Adam. https://www.geekwire.com/2017/amazon-patent-self-driving-reversible-lanes/#:~:text=The%20patent%2C%20issued%20today%2C%20concentrates%20on%20how,contact%20with%20a%20central%20roadway%20management%20system. (Accessed October 29, 2025)

I've had clients and potential clients who began developing their blockchain enterprise solutions on Amazon Managed Blockchain. They chose Amazon Web Services' Blockchain because it is faster and cheaper than building their own from scratch.

2019-NESTLÉ

In July 2019, Nestlé, a global food and beverage company, implemented a blockchain solution to enhance the transparency and traceability of its supply chain. Nestlé's blockchain enables its customers to track its products from farms to stores, ensuring authenticity and quality. In 2020, the company expanded its blockchain traceability to include a blend of arabica coffee beans from three different countries: Brazil, Rwanda, and Colombia. Consumers can now trace their coffee back to its country of origin.[184]

I used Nestle-Nespresso.com to follow Organic Kahawa Ya Congo Coffee from the field to the cup. Even though I am not much of a coffee drinker, I was impressed with the level of detail and coordination that went into making a good cup of coffee. Here are the major stops starting with South Kivu, Democratic Republic of Congo, and ending up in the US.

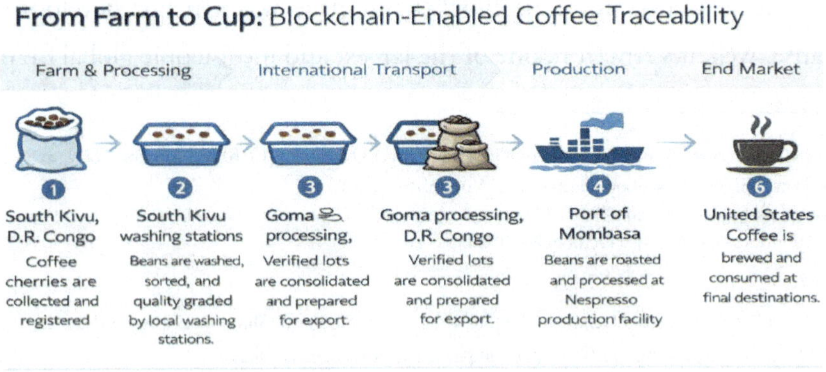

---

[184] Nestle. "Nestle Expands Blockchain to Zoégas Coffee Brand." April 6, 2020. https://www.nestle.com/media/news/nestle-blockchain-zoegas-coffee-brand (Accessed October 28, 2025)

Before researching the Nestlé blockchain, I never appreciated what it takes to make a good cup of coffee. The transparency of the Nestlé blockchain allowed me to track their coffee beans as they made 8 stops in 5 countries, which made me appreciate the complexity of the supply chain.

2021-BHP BILLITON

BHP Billiton, one of the world's largest mining companies, implemented a blockchain solution to track and record data related to mining operations. In December 2021, BHP Billiton closed a $30 million blockchain trade in copper concentrate with China Minmetals Corp's online platform MineHub. BHP's blockchain solution helps ensure the accuracy of records, increase efficiency, and improve sustainability practices.[185]

The BHP-China Minmetals partnership is a prime example of how blockchain technology can revolutionize the global mining industry. Blockchain technology has the potential to create a more reliable, ethical, and profitable mining sector by addressing key challenges, including transparency, efficiency, and sustainability.

2021-MICROSOFT

In 2021, Microsoft launched Microsoft Azure Confidential Ledger Service (ACLS). ACLS adds an extra layer of security and scalability to a blockchain. Microsoft markets ACLS as working well for auditing and tracking administrative operations.[186]

---

[185] Onstad, Eric. "BHP Completes First Blockchain Copper Concentrate Trade with Minmentals." December 14, 2021. https://www.reuters.com/technology/bhp-completes-first-blockchain-copper-concentrate-trade-with-minmetals-2021-12-14/ (Accessed November 1, 2025)

[186] Foley, Mary Jo, "Microsoft Takes Another Stab at a Blockchain-Powered Ledger Service." ZD Net, May 25, 2021. https://www.zdnet.com/finance/blockchain/microsoft-takes-another-stab-at-a-blockchain-powered-ledger-service/ (Accessed September 3, 2025)

2022-MASTERCARD

In 2022, Mastercard, a multinational payment card services corporation, launched its Cryptocard program. As of 2023, Mastercard's Cryptocard program has three billion cards in circulation worldwide and is accepted at over 100 million locations. Mastercard also offers cryptocurrency consulting services to financial institutions, cryptocurrency companies, governments, and merchants on digital currency capabilities and blockchain technologies.[187]

I use Mastercard's blockchain adoption and partnership with cryptocurrency companies as a case study in my blockchain training. The Mastercard story exemplifies the benefits of pairing credit cards with blockchain technology, enabling Web3 wallet users to access their self-custody crypto balances. If you don't use cryptocurrencies because it takes too long to process transactions or because cryptocurrencies are not accepted in most establishments, the Cryptocard program resolves both of these issues. The Mastercard Cryptocard lets you use your cryptocurrencies to purchase anywhere Mastercard is accepted.

SUMMARY

These are just a few examples of corporations that have built blockchain technology-based products and services. While these cases suggest the increasing adoption and proliferation of blockchain solutions, it is essential to note that companies utilizing blockchain technology are the exception rather than the norm at present. However, you might also remember the early 2000s, when owning and managing a website was the exception, too. Two

---

[187] Dhamodharan, Raj. "How to Turn Cryptocurrencies Into An Everyday Way to Pay." October 12, 2022. https://www.mastercard.com/news/perspectives/2022/how-turn-cryptocurrencies-into-everyday-way-to-pay/?cmp=2023.q1.glo.glo.all.brand.prod.others.mastercard-news.458701.sep.txt.googles.crypto-payment https://www.zdnet.com/finance/blockchain/microsoft-takes-another-stab-at-a-blockchain-powered-ledger-service/ (Accessed September 3, 2025)

decades later, every company not only has a website but also employs search engine optimization and maintains at least one social media page. Similarly, I believe that twenty years from now, blockchain technology will be the norm rather than the exception.

GOVERNMENTS ADOPTING BLOCKCHAIN TECHNOLOGY

Governmental adoption of blockchain technology is another significant update. Like corporations, governments are also developing blockchain solutions to enhance efficiency and improve access to government services. The government-wide adoption of blockchain technology is a positive sign that the technology is ready for use by the average citizen. [188]

This section will cover actions that governments around the world are taking to implement blockchain technology solutions. It is essential to note that in the previous chapter, we discussed global updates, with a strict focus on cryptocurrencies, CBDCs, and financial instruments. Government blockchain technology updates focus on the adoption of the technology to create innovative solutions, advance industries, and foster more effective private-public partnerships.

Listed in chronological order, the following governments have already implemented blockchain technology solutions:

2014 - ESTONIA

The Estonian government has been a leader in blockchain adoption, becoming the first country to implement blockchain technology as part of its e-Residency program. Estonia launched the e-Residency program in 2014, allowing individuals from anywhere in the world to become virtual residents

---

[188] United States Government Accountability Office Report to Congressional Requestors. Blockchain: Emerging Technology Offers Benefits for Some Applications but Faces Challenges. GAO.gov, March 2022. https://www.gao.gov/assets/gao-22-104625.pdf (Accessed October 30, 2025)

of Estonia. E-Residency uses blockchain technology to provide a secure data experience, enabling e-residents to establish and manage businesses online.[189]

The Estonian government also used blockchain technology to facilitate digital identity, land registry, and healthcare records.[190] Here are some notable blockchain technology projects in Estonia:

1. Estonia explored using blockchain technology to create a transparent and immutable registry of land ownership, thereby reducing the risk of fraud and disputes.[191]
2. Estonia also launched X-Road, a blockchain-based data exchange platform that connects various government databases, enabling government agencies to share information securely.
3. The Keyless Signature Infrastructure (KSI) is another Estonian blockchain project that secures government data and ensures the integrity of electronic records. KSI provides a tamper-proof audit trail for sensitive data, including healthcare records and legal documents.[192]

Finally, Estonia integrated blockchain technology into its healthcare system. The Estonian blockchain healthcare solution enables patients to securely access and control their health records, while also allowing healthcare providers to securely share medical data for improved coordination and patient care.[193]

---

[189] Republic of Estonia. E-Resident.gov.ee. https://www.e-resident.gov.ee (Accessed October 28, 2025)

[190] Republic of Estonia. E-Resident.gov.ee. https://e-estonia.com/solutions/estonian-e-identity/e-residency/ (Accessed October 28, 2025)

[191] Ibid.

[192] Ibid.

[193] Ibid.

UNITED ARAB EMIRATES (UAE)

The UAE has been a leader in blockchain adoption with its government-backed Smart Dubai initiative. The UAE is using blockchain technology to digitize government services and create a secure digital identity system.

Launched in 2014, the Smart Dubai Initiative (SDI) is designed to transform Dubai into a smart city and a blockchain global leader. The SDI focuses on leveraging emerging technologies, data analysis, and innovation to enhance Dubai's quality of life, economic competitiveness, and sustainability. Six key dimensions guide the SDI:

1. Smart Economy stimulates entrepreneurship, attracts investment, and fosters innovation by creating a cooperative environment for businesses. It includes initiatives such as the Dubai Data Initiative, which facilitates the sharing and use of government data to drive economic growth.
2. Smart Living improves the well-being and happiness of residents by deploying smart services. Examples include smart mobility systems, e-government services, and the Dubai Now app. I downloaded the Dubai Now app to explore its features and was genuinely impressed! It offers access to over 130 Dubai services right at your fingertips, making everything so convenient and easy to access:

    - Pay all of your bills with one click
    - Manage everything related to your vehicle
    - Sponsor your family and friends
    - Track everything in one place

3. Smart Governance uses blockchain technology to enhance government efficiency by securing government transactions.
4. Smart Environment provides efficient resource management. The Dubai Green Mobility Initiative encourages the use of electric and hybrid

vehicles, while the Dubai Integrated Energy Strategy encourages sustainable energy consumption.

5. Smart People provides individuals with the tools needed to thrive in the digital age. The Dubai Data Science Training Program and the Dubai Future Accelerators contribute to building a skilled workforce and fostering innovation.
6. Smart Mobility improves Dubai's transportation infrastructure and mobility solutions. The Dubai Autonomous Transportation Strategy aims to enhance public transportation and introduce autonomous vehicles to improve traffic management and overall transportation efficiency.[194]

The Smart Dubai initiative app plays a critical role in bridging the gap between citizens and government services. By offering a convenient, informative, and user-friendly platform, the app can empower users to actively participate in Dubai's smart city transformation. I often use the Smart Dubai app as an example of a well-constructed user interface.

All of the Dubai Now functions are very attractive and show great promise for user adoption and user interface. However, scaling to volume is one of the primary challenges that all popular apps face. The question remains: if all Dubai residents start using Dubai Now, can the app handle the increased volume?

2016 - GEORGIA

In 2016, Georgia began registering land titles and property transactions using a blockchain solution. The Georgia National Agency of Public Registry

---

[194] Blog. The Smart City in Six Dimensions. Liège University. Smart-city.uleige.be, July 2021. https://www.smart-city.uliege.be/cms/c_6946640/en/the-smart-city-in-6-dimensions#:~:text=A%20management%20that%20is%20therefore,its%20citizens'%20quality%20of%20life. (Accessed October 28, 2025)

partnered with Bitfury to design a Layer-2 private blockchain. Georgia's blockchain solutions are of special interest to legal and real estate groups.[195]

SINGAPORE

In 2016, the Singaporean government launched Project Ubin to enhance the efficiency and security of financial transactions. Project Ubin was designed to streamline the payment and settlement processes for interbank transactions and cross-border payments. It reduced processing times, enhanced transparency, and lowered costs associated with traditional clearing and settlement systems.[196] Project Ubin collaborated with financial institutions, technology companies, and other relevant stakeholders to gather insights and expertise in blockchain technology.[197]

2017 - UNITED STATES

The US government has been researching the use of blockchain technology, which encompasses a wide range of applications. Two of the completed blockchain technology solutions are explored below.

The Food and Drug Administration (FDA), specifically the Center for Drug Evaluation and Research, implemented the Real-time Application for Portable Interactive Devices (RAPID) as a blockchain-based solution to

---

[195] Snip, Inge. "Georgia: Authorities Use Blockchain Technology for Developing Land Registry." Eurasianet, April 19, 2017. https://eurasianet.org/georgia-authorities-use-blockchain-technology-for-developing-land-registry (Accessed September 3, 2025)

[196] Project Ubin: Central Bank Digital Money using Distributed Ledger Technology. Monetary Authority of Singapore. https://www.mas.gov.sg/schemes-and-initiatives/project-ubin#:~:text=What%20It%20Is,connectivity%20with%20other%20central%20banks. (Accessed September 3, 2025)

[197] Leong, Jesper. What is Project Ubin. Coinpasar.sg, March 2, 2022. https://coinpasar.sg/what-is-project-ubin/#:~:text=Project%20Ubin%20was%20a%20Singapore,companies%20and%20institutions%20in%20Singapore (Accessed October 28, 2025)

track data from the 2017 H1N1 swine flu outbreak.[198] Instead of having to read thousands of pages of information on drug trials and data from healthcare providers and hospitals, the RAPID blockchain allowed the FDA to "acquire automatically encrypted data, in real time, from many sources, and have an absolute record for all of the participants".

The Department of Health and Human Services (HHS) launched HHS Accelerate, a blockchain/distributed ledger solution to improve government efficiency and purchasing processes. The purpose was to "increase the agency's bulk purchasing power, reduce its reporting burden, cut costs for industry partners, and improve record keeping and vendor transactions."[199]

The FDA and HHS's adoption of blockchain technology involves two industries that, in my opinion, will benefit most from it: health care and government contracting. The government's healthcare and contracting procedures are complex, time-consuming, and costly. Both of these blockchain solutions successfully completed their pilot programs, optimizing internal processes and paving the way for other agencies to use blockchain technology to enhance their operations.

RUSSIA

In 2014, the Moscow government launched Active Citizen on the Ethereum blockchain. By 2017, the Moscow Department of Information Technology began using blockchain technology through the Active Citizen project, which empowers citizens to take an active role in shaping government

---

[198] Clinical Data Extraction During Public Health Emergencies: A Blockchain Technology Assessment. Public Med Central, August 30, 2021. https://pmc.ncbi.nlm.nih.gov/articles/PMC8657842/ (Accessed November 3, 2025)

[199] Ibid.

decisions by allowing them to weigh in on issues and respond to polls on the Active Citizen blockchain.[200]

2019 - SWEDEN

The Swedish government explored adopting blockchain solutions across sectors such as public administration, supply chain management, healthcare, and finance. In 2019, Sweden's land-ownership authority, Lantmäteriet, completed a trial of a blockchain-based land registration system. The system increases the security and efficiency of land transactions.[201]

2020 - CHINA

In April 2020, China launched the Blockchain-based Service Network (BSN). The BSN is a collaborative effort between the Chinese government and several major organizations, including the State Information Center (SIC), China Mobile, China UnionPay, and Red Date Technology. The BSN collaboration ensures support from government agencies, telecommunications companies, and other key stakeholders.[202]

The BSN reduces the costs of developing and deploying blockchain applications by providing shared infrastructure and resources. It is designed to facilitate the adoption of blockchain technology by enterprises, offering a standardized, open platform that reduces technical and economic barriers for business developers.

---

[200] Das, Samburaj. "Moscow to Use Blockchain Tech in 'Active Citizen' Project" March 4, 2021. CCN, https://www.ccn.com/moscow-to-use-blockchain-in-active-citizen-project/ (Accessed January 26, 2026)

[201] Camoir, John. "Sweden Conducts Trials of a Blockchain Smart Contracts Technology for Land Registry." Bitcoin Magazine, June 23, 2016. https://bitcoinmagazine.com/technical/sweden-conducts-trials-of-a-blockchain-smart-contracts-technology-for-land-registry-1466703935 (Accessed September 3, 2025)

[202] Rand, Mikk. Knowledge Base: Blockchain Based Service Network. DigiChina. DigiChina.Stanford.edu, Juky 2, 2021. https://digichina.stanford.edu/work/knowledge-base-blockchain-based-service-network-bsn-区块链服务网络/ (Accessed October 28, 2025)

BSN is important because of the collaborative effort between the Chinese government and the private sector. I believe that this type of collaboration is the best way to develop and implement a country-wide blockchain solution. The government can leverage the private sector's resources, and the private sector can provide assurance that the blockchain will comply with government regulations.

CHALLENGES IN IMPLEMENTING ENTERPRISE BLOCKCHAIN SOLUTIONS

Implementing a blockchain solution can be expensive and resource-intensive. In fact, my experience has shown that most blockchain start-ups fail due to underfunding, under-resourcing, or both. The last part of this chapter will address the reality I experience every day: the vast majority of enterprise blockchain solutions fail.

The Forbes article, "Why Enterprise Blockchain Projects Fail," details the reasons why 90% of companies and entrepreneurs fail to successfully implement blockchain solutions[203]. Reasons for high failure rates include:

1. **Lack of Clear Use Case:** One of the primary reasons for failure is a solution-in-search-of-a-problem approach. Companies may invest in blockchain due to hype, without a clear, defined business problem that the technology is uniquely suited to solve. If a traditional database or centralized system can do the job more efficiently, a decentralized ledger is often expensive to implement, unnecessary, and overly complex.
2. **Scalability and Performance:** Many of the most well-known public blockchains, like Bitcoin and Ethereum, are not built for the high transaction volumes required by enterprises. While private and

---

[203] Disparte. Dante Alighieri. Why Enterprise Blockchain Projects Fail. Forbes, May 20, 2019. https://www.forbes.com/sites/dantedisparte/2019/05/20/why-enterprise-blockchain-projects-fail/ (Accessed September 5, 2025)

permissioned blockchains offer better performance, they can still encounter scalability issues, which may result in high transaction costs and slow processing speeds.

3. **Consortium and Governance Issues:** Enterprise blockchain solutions often require a consortium of competing companies to agree on shared rules and standards. This collaboration can be difficult to manage, as organizations may be hesitant to share sensitive data or invest in a platform that doesn't provide equal benefits to all participants. The failure of projects like Maersk's TradeLens is a notable example of a consortium's inability to achieve critical mass and buy-in.

The Frontiers article, "Exploring the Failure Factors of Blockchain Adopting Projects: A Case Study of Tradelens Through the Lens of Commons Theory," provides additional insights into why enterprise blockchain solutions often fail. [204]

1. **Regulatory and Legal Uncertainty:** The evolving, often unclear regulatory landscape for blockchain and digital assets poses significant risks for businesses. Issues related to data privacy, such as GDPR's "right to be forgotten," can conflict with the immutable nature of a blockchain ledger, creating legal and compliance challenges.
2. **Complex Compliance Requirements:** In addition to the GDPR's "right to be forgotten," companies must comply with numerous regulatory requirements, including identifying data controllers and managing cross-border data transfers. These two factors create significant legal and compliance risks that often lead to project failure.
3. **Data Controller Identification:** In a traditional centralized system, the data controller—the entity responsible for data privacy and regulatory

[204] Najati, Issam. Exploring the Failure Factors of Blockchain Adopting Projects: A Case Study of Tradelens Through the Lens of Common Theory. Frontiers. Frontiersin.org, March 6, 2025. https://www.frontiersin.org/journals/blockchain/articles/10.3389/fbloc.2025.1503595/full#B49 (Accessed October 28, 2025)

compliance—is clearly defined. However, a decentralized blockchain network with nodes operated by various entities worldwide makes this assignment impossible. Under regulations like the GDPR, the "right to be forgotten" requires personal data to be erased upon request. This directly conflicts with a blockchain's core principle of immutability. Without a single, legally identifiable data controller to handle such requests, the entire project can be deemed non-compliant and is at risk of severe penalties, potentially leading to a shutdown.

4. **Cross-border Data Transfers:** The borderless nature of blockchain creates complex jurisdictional issues. When a single transaction is validated by nodes in different countries, it can be subject to multiple, often contradictory, national data laws. For example, a company operating on a global blockchain may have its data stored on a node in a country with strict data localization laws, while simultaneously being subject to the data privacy rules of the country where the transaction originated. Navigating this web of conflicting regulations is a complex challenge that raises intricate legal issues.

The Forbes and Frontier articles provide very sound reasons why blockchain projects fail, and I fully agree with their analysis. In addition to the reasons stated in the above-referenced articles, my experience is that blockchain projects mostly fail due to a lack of funding and resources. This deficiency is often symptomatic of deeper organizational and legal immaturity rather than simple budget depletion. Most technical failures stem from inadequate runway to address critical, non-code elements, such as securing compliance with evolving global regulations, performing necessary security audits (which are expensive and specialized), establishing sophisticated compliance programs, and navigating the complex legal frameworks surrounding decentralized finance (DeFi) or token issuance.

SUMMARY

In summary, while enterprise blockchain holds immense promise for enhancing transparency, efficiency, and security, the journey from proof of concept to a successful, scalable solution is fraught with obstacles. These high failure rates underscore the need for a strategic and deliberate approach that prioritizes a clear business case, robust governance, and a thorough understanding of the technology's limitations.

Blockchain technology updates encompass not only code, programming, and protocol updates, but also span a wide range of industries and regulatory perspectives, including governmental adoption and the development of new blockchain solutions. As the updates in this chapter demonstrate, many countries and Fortune 500 companies have not only adopted this technology but are also actively fostering innovation and providing blockchain products and services.

Blockchain technology is still relatively new, but it is already undergoing significant evolution. This chapter presented compelling examples of corporate and government efforts to adopt blockchain technology.

The following chart summarizes government blockchain solutions by country and industry.

Country	Land Registry	Health Care	Government Data	Financial	Residency	Business Support	Supply Chain
China						x	
Estonia	x	x	x		x		
Georgia	x						
Russia	x						
Singapore				x			

Sweden	x	x		x			x
UAE							
USA		x	x			x	

The Dubai Smart Cities App is an excellent example of a government blockchain initiative that is accessible worldwide and features a user-friendly interface. I downloaded the Dubai app in minutes and was able to explore its features seamlessly, significantly improving my overall experience. This app is a great example of user adoption and interface design.

SUCCESS STRATEGY ACTION ITEMS

1. As Ethereum is the company that established the integration of blockchain technology in the business environment, I always advise investors to keep track of Ethereum updates.

2. Research and follow blockchain updates from corporations listed in this chapter, as well as other companies. The blockchain industry is still rapidly evolving, so a lot can happen in a month or even a week. If you want to stay informed about the latest advances in blockchain technology and investment opportunities, you need a process to stay up to date on blockchain updates.
3. The Estonian e-residency program is a good example of a government-based blockchain solution. While the benefits of e-residency are not as comprehensive as those of standard residency, the program offers individuals the option to apply for Estonian residency.

4. The Start Dubai Initiative app is another great example of a government-based blockchain application with a user-friendly interface. Global government agencies and contractors can use the Smart Dubai Initiative App as a benchmark for government blockchain apps with a robust public-private user interface. Anyone can download the Smart Dubai App to review all its functions.

CHAPTER 7
THE A-Z OF BUYING NFTS

"NFTs are transforming the way commerce is transacted, contracts are enforced, investments are held, and value is transferred."[205]

Hendrith Vanlon Smith Jr.
Managing Partner and CEO, Mayflower – Plymouth Capital, LLC.

CHAPTER OVERVIEW

Non-fungible tokens (NFTs) have emerged as a groundbreaking innovation in blockchain technology, revolutionizing the way we think about ownership and value. Simply put, NFTs are unique digital assets that certify ownership of a specific item or asset. NFTs have expanded the range of assets that can be tokenized on a blockchain. Beyond traditional cryptocurrencies like Bitcoin and Ethereum, NFTs have opened exciting new doors by allowing the tokenization of digital art, music, collectibles, gaming, and even real-world assets, such as real estate. This innovation is creating new ways for people to own, trade, and monetize digital and physical assets, making things more accessible and engaging.

At the mention of NFTs, the average person is bound to think about art, music, or gaming. However, observe that the Hendrith Vanlon Smith Jr.

[205] Quote Fancy. QuoteFamcy.com. https://quotefancy.com/quote/3732440/Hendrith-Vanlon-Smith-Jr-NFT-s-are-transforming-the-way-commerce-is-transacted-contracts (Accessed September 24, 2025)

quote above does not mention transforming art, music, or gaming. Instead, it explicitly describes NFTs as "transforming the way commerce is transacted." As you will learn in this chapter, there is so much more to NFTs than art, music, and gaming.

NFTs have been used in commerce to record and transfer event tickets and educational certificates. NFTs have been used to transfer real estate titles[206]. The interesting and often underappreciated part of the above quote is the reference to "value." How do NFTs transact value? One example, which is covered later in this chapter, is voting. NFTs could be used to record and validate votes in any jurisdiction worldwide that utilizes blockchain technology to track and verify election results. For example, West Virginia was the first US state to use blockchain technology in a national election.[207] Votes have immense value and are the cornerstone of every democracy.

Like cryptocurrencies, NFTs experienced a significant price increase, followed by an equally substantial sell-off. Unlike cryptocurrencies, the overall NFT market never fully recovered from its losses, let alone reached new highs. However, the wealth in NFTs lies in creating them, not purchasing them. This chapter delves into the process of buying NFTs.

MY NFT JOURNEY

My first encounter with NFTs was quite fascinating. I first heard about NFTs and OpenSea.io in 2019, and one weekend, I decided to open an account on 'OpenSea.io' and explore the world of NFTs. As I browsed through the collection on 'OpenSea.io', I immediately recognized the album

---

[206] Samkian, Kevin. NFTs: The Future of Real Estate? Pepperdine Caruso School of Law, January 17, 2023. https://law.pepperdine.edu/surf-report/posts/nfts-the-future-of-real-estate-kevin-samkian.htm (Accessed September 4, 2025)

[207] Miller, Ben. West Viginia Becomes First State to Test Mobile Voting by Blockchain in a Federal Election. GovTech.com. https://www.govtech.com/biz/west-virginia-becomes-first-state-to-test-mobile-voting-by-blockchain-in-a-federal-election.html (Accessed September 3, 2025)

cover for Notorious B.I.G.'s Greatest Hits. It shows the iconic picture of Notorious B.I.G.'s head from the nose up, wearing a Kangol hat. The NFT image of his face and the Kangol were pixelated, and the overall look reminded me of a mosaic made of tiny boxes, but it was clear to me that it was the album cover.

As an attorney with experience in intellectual property law, my first thought was, "This has to be trademark infringement!" However, I decided to gather more information before reaching a legal conclusion. As it turns out, my legal conclusion was probably correct. I say "probably" because NFT legal issues are just starting to work their way through the court system. The question of whether a pixelated NFT infringes on a trademark has yet to be decided by a court. However, this initial experience exemplifies the surprising possibilities and the uncertainties associated with NFTs, as we shall discuss in the following pages.

NFT DEFINITIONS

There are numerous NFT definitions. To gain a fuller understanding, let's compare three NFT definitions.

> *Wikipedia*: A unique digital identifier that is recorded on a blockchain and is used to certify ownership and authenticity. It cannot be copied, substituted, or subdivided. The ownership of an NFT is recorded in the blockchain and can be transferred by the owner, allowing NFTs to be sold and traded.[208]
>
> *Merriam-Webster*: A unique digital identifier that cannot be copied, substituted, or subdivided, that is recorded in a blockchain, and is used

---

[208] Non-Fungible Token. https://en.wikipedia.org/wiki/Non-fungible_token Wikipedia, (Accessed September 29, 2025)

to certify authenticity and ownership (as of a specific digital asset and specific rights relating to it).[209]

CoinMarketCap: A non-fungible token (NFT) is a type of digital asset that represents ownership or proof of authenticity of a unique item or piece of content, utilizing blockchain technology. Unlike cryptocurrencies like Bitcoin or Ethereum, which are fungible and can be exchanged on a one-to-one basis, each NFT is distinct and cannot be exchanged on a like-for-like basis with another NFT. This uniqueness makes NFTs suitable for representing individual pieces of art, collectibles, music, videos, and more.[210]

These three NFT definitions are fairly consistent, with one difference. Unlike the definitions from Wikipedia and Merriam-Webster, the CoinMarketCap definition does not state that an NFT cannot be subdivided. For the purposes of this book, I call NFTs that cannot be subdivided 'traditional NFTs.' Later versions of NFTs allowed them to be divided into smaller units commonly referred to as a 'fractionalized NFT.'

Fractionalized NFTs are NFTs divided into smaller fractions that can be owned and traded by multiple people. They are also referred to as "wrapped" NFTs, which enable broader participation in the ownership of high-value or exclusive digital assets. For example, expensive NFTs could be tokenized and fractionalized, and people could buy and own smaller portions of that

---

[209] NFT. https://www.merriam-webster.com/dictionary/NFT#:~:text=,en%2D(,)ef%2D'tē,specific%20rights%20relating%20to%20it) (Accessed September 3, 2025)
[210] Non-Fungible Token. Coinmarketcap.com. https://coinmarketcap.com/academy/glossary/non-fungible-token (Accessed December 1, 2025)

artwork's NFT, effectively dividing ownership. However, the owner of the NFT does not own any of the actual art.[211]

DEFINITIONS OF NFT TERMS

For clarity and ease of reference, this section will cover the definitions of the most common NFT terms in alphabetical order:

- **Air Drops**: New NFTs that are automatically sent to your wallet for free. This is a common practice to attract and reward early adopters of projects.
- **Fractionalized Ownership**: Partial NFT ownership rights. Sellers can sell a portion of an NFT, and buyers can buy a portion of an NFT.
- **GameFI**: (Gaming and Decentralized Finance) Decentralized applications (dApps) that mint reward tokens or NFTs when players play the game.
- **Pump and Dump**: A type of scam. A pump occurs when a person or a group of 'whales' buys or convinces others to purchase large quantities of a cryptocurrency or an NFT to increase prices. When the price peaks, these individuals quickly exit their position at the peak to make a substantial profit, which in turn drives the cost down. Investors who bought in at the peak price usually lose most or all of their investment.
- **Rarities:** Digital collectibles and in-game NFTs that typically come with different rarities (such as common, uncommon, rare, legendary, and mystic) and get assigned a certain edition number (e.g., #001/100).
- **Reveal:** When you mint NFTs, especially for new generative projects, the artwork won't actually be created until the minting is complete. This

---

[211] Copeland, Tim. "What are Fractional NFTs?" The Block, September 28, 2023. https://www.theblock.co/learn/251480/how-fractionalization-works-in-the-nft-market (Accessed October 28, 2025)

means you're purchasing without knowing what you'll get until the reveal.

- **Roadmap:** The set of activities or development plans for NFT projects or P2E games.
- **Rug pull:** A scam in which the team behind a seemingly legit NFT project disappears with all the money raised immediately after launch, making the actual NFT or the tokens you purchased worthless.
- **Snap:** Buying an NFT quickly and for a very low price, also known as "sniping". [212]

Understanding these NFT terms will help you understand the NFT market. All of these terms will be used liberally in this chapter, so having a basic understanding will help you dive deeper with more ease.

The best way to describe how NFTs function is to compare a traditional transaction using fiat currencies to an NFT transaction. For the purposes of this scenario, we will compare an NFT transaction to a fiat currency transaction, highlighting the differences between them in five key categories: ownership and physicality, fungibility, tangibility, portability, and divisibility.

NFT SCENARIO: THE ART AUCTION

During an art auction, two individuals, Fiat Francis and NFT Nick, are interested in purchasing art. Francis is purchasing a traditional painting, while Nick is purchasing an NFT representing a digital artwork. This scenario is designed to highlight the fundamental differences between fiat currency and NFTs.

[212] Non-Fungible Token. https://en.wikipedia.org/wiki/Non-fungible_token (Accessed September 2, 2025)

1. Ownership and Physicality:

Francis: She wins the bid for a traditional painting and pays with fiat currency. She receives the physical painting, which she can hang on her wall.

Nick: He wins the bid for the NFT, but instead of receiving a physical item, he gains ownership of a unique digital token stored on a blockchain. This proves his ownership of the NFT, but he does not get to own the actual, physical artwork.

2. Fungibility:

Francis: Her fiat currency is fungible, meaning each dollar bill of the same denomination is interchangeable with another and holds the same value.

Nick: His NFT is non-fungible, meaning it is unique and not interchangeable with other NFTs. Each NFT represents a distinct piece of digital art, and its value can vary based on its rarity and demand.

3. Tangibility:

Francis: She physically holds her fiat currency, which she can use for various purposes, such as buying goods, paying bills, or investing in other assets.

Nick: His NFT is digital, so he can't physically touch or display the artwork, but he can showcase his ownership through the blockchain.

4. Portability:

Francis: She can use her fiat currency for transactions wherever cash is accepted.

Nick: His NFT is in a digital wallet, which allows him to transfer or sell it online to anyone with blockchain access.

5. Divisibility:

Francis: She can divide her fiat currency into smaller denominations to make change for various transactions.

Nick: Traditional NFTs are not divisible. Each NFT represents a whole, unique digital item, and you can't divide it into smaller parts, unlike fiat currency.

This example highlights the key differences between fiat currencies and NFTs. While fiat currencies are physical, interchangeable, and accepted for many transactions, NFTs are digital, non-interchangeable, and represent ownership of unique digital or physical items. These differences influence how NFTs are used, stored, and traded in commerce.

A BRIEF HISTORY OF NFTS

Although the NFT craze emerged in 2020, these digital assets have been around since 2014, when they were not yet referred to as NFTs.

2014 - QUANTUM

A version of the first NFT ever created was Kevin McCoy's "Quantum" (2014-2021). Quantum sold for $1.4 million, including fees, at the Sotheby's auction Natively Digital: A Curated NFT Sale. The buyer, known as "Sillytuna," also sold a "CryptoPunk" NFT from his collection in the same auction for $11.8 million.[213]

Supply and demand are always price factors, and since there was only one Quantum, I believe the $1.4 million price tag was based on its extreme rarity. If 100, or even 10, versions of Quantum were up for auction, they would not

---

[213] Di Liscia, Valentina. First Ever NFT Sells for $1.4 Million. Hyperallergic.com, June 10, 2021. https://hyperallergic.com/652671/kevin-mccoy-quantum-first-nft-created-sells-at-sothebys-for-over-one-million/ (Accessed October 29, 2025)

have sold for $1.4 million each. Moreover, the sale of Quantum at Sotheby's, a prestigious auction house, further elevated interest in Quantum, its brand, and, ultimately, its final auction price.

2017 - CRYPTOPUNKS

CryptoPunks are NFTs launched by Larva Labs in 2017 on the Ethereum blockchain, specifically using the ERC-20 token standard. Each CryptoPunk is a unique, pixelated character with distinct attributes. This scarcity and uniqueness created a sense of exclusivity and desirability among collectors. No two CryptoPunks are the same, and there are only 10,000 pieces in total. One CryptoPunk NFT even reached the $530 million mark, making it the highest NFT sale to date.[214]

As one of the first NFT projects on the Ethereum block-chain, CryptoPunks benefited from early adoption and a significant first-mover advantage. They established themselves as a pioneer in the NFT space, capturing the attention of early adopters, collectors, and investors.

CryptoPunks gained significant visibility through endorsements from celebrities and influencers. For example, the following celebrities purchased CryptoPunks:

- **Jay-Z:** Rapper and billionaire businessman.
- **Odell Beckham Jr.:** Star college and NFL football player.
- **Serena Williams:** The former #1 tennis player in the world was actually given her CryptoPunk as a gift by her husband, Alexis Ohanian, Co-Founder of Reddit.[215]

---

[214] Van Boom, Daniel. "Why This CryptoPunk NFT Sold for $532 Million. Sort of." CNET, December 19, 2021. https://www.cnet.com/personal-finance/crypto/why-this-cryptopunk-nft-sold-for-532-million-sort-of/ (Accessed October 28, 2025)

[215] Ginsburg, Randy. Celebrities Who Own CryptoPunks: 15 Famous NFT Owners. NFT Now, April 11, 2022, https://nftnow.com/collectibles/from-celebrities-to-ceos-meet-15-of-the-most-famous-cryptopunk-holders/ (Accessed November 1, 2025)

These celebrities helped legitimize CryptoPunks and NFTs in the public eye, which attracted new buyers to the market. CryptoPunks also has a strong and engaged community that has played a crucial role in driving demand and supporting the project's long-term success.

These factors combined to make CryptoPunks a cultural phenomenon and a valuable asset for many collectors. While its success has inspired numerous other NFT projects, CryptoPunks remains a benchmark for the NFT space due to its early adoption, scarcity, and strong community.

I did not learn about CryptoPunks until 2018. When I learned about CryptoPunks, I began paper trading them. I tracked the CryptoPunks' prices for a year, and they continued to increase in value. And despite their rising prices, I did not find them worth my investment. From my perspective, an investment must have a clear business purpose and address a specific problem. That was not the CryptoPunk business model, so I was never tempted to purchase any CryptoPunks. I did consider the alternative of purchasing CryptoPunks as a work of art, but I was not interested in digital art at the time.

Since 2022, CryptoPunks have experienced a series of interesting events. In 2020, Yuga Labs, creators of The Bored Ape Yacht Club, purchased CryptoPunks during the peak of the crypto bull run. Since the crypto crash, most cryptocurrencies have recovered, but NFTs have not recovered at the same rate. So, in 2025, Yuga Labs sold CryptoPunks to the non-profit Node Foundation, which focuses on preserving digital art.[216] The sale of CryptoPunks to a non-profit is a strong signal that NFTs may never reach the record highs of 2021 again.

---

[216] Davis, Dominic-Madori. NFT Phenom CryptoPunks Was Just Sold as to a Non-Profit. TechCrunch.com. May 13, 2025. https://techcrunch.com/2025/05/13/nft-phenom-cryptopunks-was-just-sold-to-a-nonprofit/ (Accessed October 28, 2025)

CRYPTOKITTIES

CryptoKitties is a blockchain-based virtual game that allows players to adopt, breed, and trade virtual cats. CryptoKitties was created by Axiom Zen, a Vancouver-based company, and was introduced during the world's largest hackathon for the Ethereum ecosystem. The transaction activity was so high that CryptoKitties clogged the Ethereum blockchain, which made it even more prominent.[217] CryptoKitties' success can be attributed to several factors:

- CryptoKitties introduced a unique concept of collectible digital cats that could be bred, traded, and owned on the blockchain. This gamification element appealed to a wide audience, particularly because users could earn cryptocurrency by breeding and trading CryptoKitties.

- The limited supply of CryptoKitties created a sense of scarcity, making them desirable to collectors. Each cat had unique attributes and traits, further enhancing their value.

- A strong and engaged community of CryptoKitties enthusiasts played a crucial role in the project's success. The novelty and popularity of CryptoKitties attracted significant media attention, which helped gain traction and drive user adoption.

I was introduced to CryptoKitties during a CryptoKitties presentation at a 2017 blockchain event. I recall thinking that CryptoKitties wouldn't be popular or successful. Months later, CryptoKitties experienced a surge in value. Then CryptoKitties made a significant change.

---

[217] Cryptopedia Staff. CryptoKitties: A Pioneer in Ethereum Gaming and NFTs. Gemini. Gemini.com. March 10, 2022. https://www.gemini.com/cryptopedia/cryptokitties-nft-crypto-ethereum-token (Accessed October 28, 2025)

CryptoKitties created their own blockchain, the Flow blockchain, and left Ethereum. But this was not the only change for CryptoKitties; it also experienced a significant price drop. Other than the 2018 spike driven by CryptoKitties' novelty, I believe part of the price drop was due to the CryptoKitties business model.

When CryptoKitties mate, one sire (male) and one dame (female) combine to create a new generation of CryptoKitties. This mating process directly affects the balance between supply and demand. As the mating process increases the supply, I believe that eventually the same process decreases the demand, which also reduces the price. CryptoKitties' price never recovered to its 2018 peak.

Cryptopunks and CryptoKitties were two of the main NFTs that sparked the NFT craze. The next major NFT was the Bored Ape Yacht Club.

2021-THE BORED APE YACHT CLUB

The BAYC, created by Yuga Labs, is an NFT on the Ethereum blockchain trading with the ERC-721 token. Most experienced NFT owners know about the Bored Ape Yacht Club (BAYC).

What is the BAYC? Why is it so successful? Should you buy a BAYC NFT? Why do celebrities buy BAYC NFTs? This section will answer these questions.

The title Bored Ape Yacht Club is a clever combination of three components: "Apeing in", "Bored" wealthy investors, and "Yacht Club".

- "Apeing in" is a crypto term that means investing aggressively without conducting research.[218] By the way, as stated numerous times in this

[218] Legge, Michelle. Apeing. Koinly.io, November 21, 2023. https://koinly.io/crypto-glossary/apeing/ (Accessed October 28, 2025)

book, investors should always conduct thorough research before making an investment.

- "Bored" refers to the perspective that investors are so rich from investing in cryptocurrencies that they are bored.[219]
- "Yacht Club" refers to the exclusive nature of buying a Bored Ape. Purchasers are buying into an exclusive club and a lifestyle.[220]

The title "Bored Ape Yacht Club" also pretty accurately describes the artistic rendering of the BAYC NFTs, a series of portraits of apes with mostly bored looks on their faces. I say "mostly" because some of the Bored Apes don't look bored — they have silly or twisted smiles on their faces.

Each Bored Ape is generated based on seven traits:

- Background color
- Style of clothing
- Earring
- Eyes
- Fur
- Hat
- Mouth

Some Bored Apes have rarer traits than others. The rarer the trait, the higher the value of the Bored Ape NFT.

BAYC launched a live presale on April 23, 2021, and the NFTs were initially sold for .08 ETH. By April 30, 2022, the BAYC NFTs were selling

---

[219] The Shrimply Team. What Are Bored Apes NFTs? Complete Guide to Bored Ape Yacht Club. Academy.shrimply.10, May 8, 2023. https://academy.shrimpy.io/post/what-are-bored-ape-nfts-complete-guide-to-bored-ape-yacht-club#:~:text=Bored%20Ape%20NFT%20Rarity%20and,of%20149%20with%20cross%20earring (Accessed October 28, 2025)

[220] Sergeenkov, Andrey. What is the Story Behind Bored Ape Yacht Club Creator Yuga Labs. CoinDesk.com, May 11, 2023. https://www.coindesk.com/learn/whats-the-story-behind-bored-ape-yacht-club-creator-yuga-labs (Accessed September 3, 2025)

for 153.70 ETH, a 192,125% increase. Then, the cryptocurrency and NFT market correction occurred on April 3, 2021, and the average BAYC NFT was selling for 62.263 ETH. The CoinGecko.com chart tells the BAYC price story.

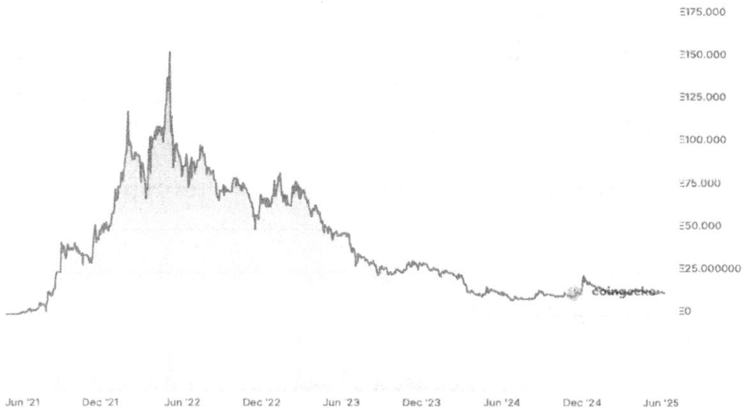

WHY DID THE BORED APE YACHT CLUB TAKE OFF?

The BAYC combined a desirable lifestyle brand and NFTs. Most BAYC club purchasers do not buy the BAYC NFT because of the creativity and uniqueness of the Bored Apes in the NFTs. Don't get me wrong, some of the BAYC NFTs are very unique — and some are even a bit disturbing — but the NFT art is only the packaging of the BAYC NFT.

The true value of the BAYC NFT lies in what is associated with it. Rather than just selling an NFT, Yuga Labs is selling a lifestyle— and people are buying! According to CoinMarketCap's article, *How Did Bored Ape Yacht Club Become the Most Popular NFT Collection to Exist – Diving into Celebrity Investments,* becoming a BAYC club member comes with access to the following perks:

- An exclusive Discord server reserved for BAYC members only.
- Randomly distributed free additional NFT collectibles that are airdropped into the wallets of members.

- A yacht club members-only ticket with undisclosed benefits. One of the benefits released was a community graffiti board called "the Bathroom" for digital art. This keeps the members engrossed in the BAYC's universe-like appearance.
- The ability to turn their apes into "mutants" by using airdropped serums freely. These mutant apes can also be sold for about 3 ETH.
- Exclusive games, such as the Mutant Arcade, added to the exclusive atmosphere that BAYC offers.
- BAYC also launched ApeCoin, which has been airdropped to Bored Ape NFT holders.[221]

SHOULD YOU BUY A BAYC NFT?

According to Cryptoclam.io, at the time of publication, the average BAYC NFTs cost $23.89K.[222] If you, like 99.99 percent of the global population, do not have that kind of money to spend, the answer is "no". If you do have that kind of money to spend, then the next question is whether the BAYC perks are worth your investment dollars.

Apart from showing off your Bored Ape on your social media pages, you must decide if it is worth spending that kind of money on the BAYC lifestyle, prestige, and access. Although the BAYC has a business model and numerous perks, I personally saw better returns on my investment dollars in cryptocurrencies.

---

221 Ahmed, Sarah. "How Did Bored Ape Yacht Club Become the Most Popular NFT Collection to Exist – Diving Into Celebrity Investments." Coinmarketcap.com, https://coinmarketcap.com/academy/article/how-did-bored-ape-yacht-club-become-the-most-popular-nft-collection-to-exist-diving-into-celebrity-investments. (Accessed September 3, 2025)

222 Bored Ape Yacht Club. DappRAdar.com. https://dappradar.com/dapp/bored-ape-yacht-club (Accessed September 4, 2025)

NEW ERA FOR NFTS

Newer versions of NFTs often have functional attachments that are integrated into the NFTs. This means that when you purchase a new-era NFT, it may have an attachment that adds additional value. For example, a new-era NFT may be a piece of art that also includes a ticket to an art gallery or a fireside chat with the artist. Another example discussed earlier in this chapter is album NFTs, which often include tickets to a live concert, additional songs, or even a virtual concert in the metaverse. Here are some of the emerging functional uses for NFTs:

- NFTs can be used for event tickets, eliminating the possibility of fraud and scalping. Because each NFT is unique and traceable on the blockchain, it can't be easily replicated or resold for a price above a certain threshold.
- NFTs can track the movement of goods through a supply chain, ensuring authenticity. This can be especially useful for items such as luxury goods or pharmaceuticals.
- NFTs can represent whole or fractional ownership of real-world assets, like real estate or artwork. This makes investment and portfolio diversification easier.
- NFTs can also be used in DeFi applications to create new financial instruments, such as loan collateral.

NFTs can be used to create a secure and verifiable digital identity, which is useful for applications such as online voting, digital government identification, or access control to specific websites, computer systems, or buildings. For voting, each voter could be assigned a unique NFT that represents their voting rights. A voting NFT would be stored on a blockchain, a secure and tamper-proof digital asset.

These are just a few examples of the new era of NFTs, but innovators will undoubtedly find new use cases for NFTs. These new use cases will continue to evolve the NFT and digital landscape.

NFTS IN POP CULTURE

Similar to cryptocurrencies, NFTs have also gained popularity in pop culture. NFT references started appearing in movies, TV shows, and music.

FUTURAMA REBOOT

I am a huge Futurama fan, so I was overjoyed when Hulu brought back the original Futurama cast for Season 8. I was even more surprised to see Futurama make fun of NFTs in Season 8 episode 1, when a young self-proclaimed "important tech guy" driving a luxury hovercar arrogantly has the following conversation with his car's AI, "Sniri":

> *Tech Guy* – (In an upbeat and arrogant voice) Sniri, make a reservation for me at the Impassable Burger restaurant and check on the price of my NFTs.
>
> *Sniri* – (in a matter-of-fact voice) The restaurant has been confirmed. NFTs worthless...

Other than the typical hilarious Futurama delivery, I found this funny because I have always maintained that 95% of NFTs are worthless, especially if they lack any intrinsic value attached to them (such as tickets, books, music, movies, etc.).

THE TONIGHT SHOW WITH JIMMY FALLON

In August 2021, on The Tonight Show, Paris Hilton educated Jimmy Fallon about NFTs. Then, Paris and Jimmy discussed Beeple's NFTs, and Paris even playfully recommended that Jimmy sell one of his jokes as an NFT. The entire conversation lasted only about 54 seconds, but it exposed a considerably wider audience to NFTs in a simple, clearly understandable way, demystifying this new technology.

In January 2022, during another appearance on The Tonight Show, Paris and Jimmy revisited NFTs, specifically discussing the Bored Ape Yacht Club.

Paris Hilton's appearances on The Tonight Show with Jimmy Fallon in 2021 were significant for two key reasons. First, Hilton's large following and celebrity status helped bring NFTs and the Bored Ape Yacht Club to a broader demographic. Second, her endorsement of NFTs added legitimacy and credibility to the NFT marketplace. Throughout this book, we've discussed in detail the significant impact of celebrity endorsements on cryptocurrency. In this case, NFTs entered the mainstream sphere through a trifecta of endorsements: Paris Hilton's fans, Jimmy Fallon's fans, and The Tonight Show's fans.

SESAME STREET

Yes, Sesame Street is selling NFTs! The beloved kids' TV show partnered with VeVe to sell 5,555 Cookie Monster NFTs priced at $60 each. Jennifer Ahearn, Sesame Workshop's Vice President of Global Strategic Partnerships and Themed Entertainment, stated that Cookie Monster NFTs "give the fans a new way to showcase their love for Sesame Street and its characters."[223] This marks a major milestone for NFTs, since Sesame Street will introduce them to a young audience that might grow up using NFTs in their everyday lives. While NFTs are a new and groundbreaking concept today, they have the potential to become second nature for the coming generations, shaped by early exposure through platforms like Sesame Street.

---

[223] White, Gemma. "Sesame Street's Cookie Monster NFT and 10 Pop Culture Moments on the Blockchain." The National News.com, March 15, 2023. https://www.thenationalnews.com/arts culture/pop culture/2023/03/15/sesame-streets-cookie-monster-nft-and-10-pop-culture-moments-on-the-blockchain/ (Accessed September 3, 2025)

NFTS IN MUSIC

Musicians and DJs have also used NFTs to expand their audience and increase profits.

American DJ Justin Blau, also known as 3LAU, was an early adopter of music NFTs. 3LAU sold his first NFTs in the fall of 2020, but that was just the beginning. In February 2021, his Ultraviolet collection of 33 different NFTs reportedly generated $11.7 million for him.[224]

In March 2021, Kings of Leon became the first musical group to release their album and NFT simultaneously. Using the YellowHeart platform, the American rock band generated 820 ETH ($1.45 million at the time). Exclusive versions of the album, unique digital artwork, and lifetime concert passes were also included in the NFT.[225]

Dutch DJ Don Diablo made NFT history in 2021 by offering the first NFT full-concert film, Destination Hexagonia, through SuperRare.[226]

Music has always been at the forefront of technology, from records to cassettes to CDs to digital music. Could NFTs be the next musical revolution? Before NFTs, music purchasers only received the music. Now, NFT listeners can receive lyrics, tickets, special invitations, backstage passes, artwork, or anything else in the artist's imagination at the same time. I believe NFTs are set to redefine the future of the music industry, transforming how artists create, share, and monetize their work.

---

224 Harris, Chris; Thomas, Langston. The Top Music NFT Moments of All Time. NFTNow.com, August, 5, 2022. https://nftnow.com/music/top-music-nft-moments/#:~:text=These%20sales%20helped%20pave%20the,the%20way%20that%20it%20did. (Accessed September 30, 2025)

225 Ibid.

226 Harris, Chris and Thomas, Langston. The Top Music NFT Moments of All Time. NFTNow.com., August 5, 2022. https://nftnow.com/music/top-music-nft-moments/ (Accessed October 28, 2025)

ORDINALS, RUNES, AND NFTS

As discussed earlier in this chapter, NFTs represent ownership of unique digital or physical assets, such as art, collectibles, or real estate. They often utilize smart contracts to automate transactions and enforce ownership rights. Ordinals and runes are two relatively new concepts in the Bitcoin ecosystem that can be considered similar to NFTs.

ORDINALS

Ordinals are essentially inscriptions on individual satoshis, the smallest unit of Bitcoin. Think of each satoshi like pennies to a dollar — one Bitcoin is equivalent to 100 million satoshis. Each satoshi can be assigned a unique inscription or an ordinal, allowing the creation of digital collectibles or other forms of data on the Bitcoin blockchain. While Bitcoin itself is fungible (one Bitcoin is interchangeable with another), ordinals can represent unique, non-fungible assets. They are native to the Bitcoin blockchain, leveraging its security and decentralization. Because the data is stored on the Bitcoin blockchain, it benefits from Bitcoin's immutability and security. SATS, trac, and rats are three examples of ordinals.

The real-world uses for Bitcoin ordinals are very similar to NFTs.

1. **Digital Art and Collectibles:** This is the most prominent use case. Artists can inscribe their digital artwork, images, GIFs, or even audio directly onto a satoshi. Unlike NFTs on other chains that might link to off-chain data, ordinals store the actual content on-chain, ensuring its permanence.
    - Examples include "Bitcoin Punks," "Bitcoin Rocks," and numerous other collections that have emerged as unique digital art series.
2. **Gaming:** Ordinals can be used to represent in-game assets, characters, or unique items within blockchain-based games.

3. **Data Anchoring/Immutable Records**: Ordinals can be used to store important data that requires tamper-proof and verifiable integrity. Examples include:
    - Legal Documents: Storing legal agreements or contracts.
    - Intellectual Property: Securing ownership and licensing information for digital and physical assets.
    - Supply Chain Management: Recording the origin and movement of goods to enhance transparency and reduce fraud.
    - Academic Credentials/Certificates: Storing verifiable proof of credentials.
4. **Proof of Ownership**: Ordinals can serve as digital certificates of ownership for real-world assets, including real estate and luxury goods.

RUNES

Runes are a fungible token protocol on Bitcoin, designed to efficiently issue and manage interchangeable tokens. Built by Casey Rodarmor (the same mind behind ordinals), Runes were designed to do one thing well: make it simple and Bitcoin-native to issue and manage interchangeable digital assets directly on Bitcoin.[227]

Here's the clean distinction. Ordinals are about uniqueness — inscriptions tied to individual satoshis. Runes are about interchangeability — more like ERC-20 tokens on Ethereum. One unit of a rune is meant to be the same as the next, which is exactly why they work for currencies, utility tokens, rewards, and other standardized assets.

What I like about the runes design is that it stays true to Bitcoin's DNA. It uses Bitcoin's existing transaction model and avoids smart contracts. That

[227] Everything You Need to Know About Bitcoin Runes: How They Work and When They Use Them. Pixelplex.io, October 14, 2024. https://pixelplex.io/blog/bitcoin-runes/ (Accessed December 1, 2025)

keeps things lean, secure, and decentralized — while still expanding Bitcoin beyond simple peer-to-peer payments.

Runes are not NFTs, but they belong in the same conversation as ordinals because they push Bitcoin into a broader digital-asset era: ordinals for non-fungible artifacts, and runes for fungible tokens. Here are some real-world uses for Bitcoin runes:

1. **Fungible Tokens**: Runes are ideal for creating interchangeable tokens, similar to ERC-20 tokens on the Ethereum blockchain. This includes:
    - Memecoins: Runes can be used to create and trade meme coins.
    - Loyalty Points/Rewards: Businesses could issue their own loyalty points or reward tokens on the Bitcoin blockchain.
    - In-game Currencies: Creating fungible currencies for games.
2. **DeFi on Bitcoin**: Runes present new DeFi opportunities, including:
    - Lending and Borrowing Platforms: Enabling users to lend or borrow Bitcoin or other tokenized assets.
    - Liquidity Pools: Facilitating decentralized trading and swapping of tokens.
    - Decentralized Exchanges (DEXs): Building platforms for peer-to-peer token trading on Bitcoin.
3. **Asset Tokenization**: Runes can also tokenize real-world assets, including:
    - Fractional ownership of real estate,
    - Commodities, or
    - Traditional financial instruments, such as stocks and bonds.
4. **Stablecoins**: Runes could be used to issue stablecoins pegged to traditional fiat currencies directly on the Bitcoin blockchain.

5. **Layer-2 Integration**: Runes are compatible with Bitcoin's Lightning Network, enabling faster, more cost-effective transactions and making Bitcoin more practical for everyday use and micropayments.

In essence, while ordinals focus on *unique* digital artifacts, runes are tailored for *interchangeable* digital assets. Both expand Bitcoin's utility beyond its original scope, attracting new users and developers to the network by enabling a wider range of decentralized applications and digital asset creation. The key factors that distinguish ordinals and runes from NFTs are as follows:

- **Blockchain**: Ordinals and runes are specifically tied to the Bitcoin blockchain. NFTs can be created on various blockchains.
- **Underlying Asset**: Ordinals are inscriptions on individual satoshis, while runes are domain names. NFTs can represent a wide range of digital or physical assets.
- **Functionality**: Ordinals and runes have specific use cases within the Bitcoin ecosystem. NFTs are more versatile and can be used for a wide range of purposes.

I view ordinals and runes as very important updates to the Bitcoin blockchain. After NFTs, originally tied to the Ethereum blockchain, became a multi-billion-dollar industry, Bitcoin had to evolve and adapt. However, NFTs have undergone significant advancements in marketing and technology over the past few years. The question remains: *Is Bitcoin too late to the NFT game?*

HOW TO GET STARTED INVESTING IN NFTS

Never invest in anything you do not understand — that's the one cardinal rule I tell all my clients and trainees. Before you start investing in NFTs, you must understand exactly how they work. My own journey of understanding

NFTs began with exploring NFT marketplaces, including Opensea, Rarible, SuperRare, and Nifty Gateway. Over the course of several weeks, I researched and analyzed over 100 NFTs, including both successful and unsuccessful ones. I also examined several NFT websites and social media pages, gathering insights into the factors that drive success or failure in this space. This was my first step towards truly understanding NFTs, so I could make wise and informed investment decisions.

If you are buying NFTs for a return on investment, you must understand and determine what makes a profitable NFT. Like stocks, investors may have different interpretations of what makes a profitable NFT. However, most profitable NFTs share some common traits, such as:

- Rarity/uniqueness
- High volume/transaction rate.
- Independent demand.
- Tells a good story.

I recommend using the same approach I did when I started learning about NFTs: familiarize yourself with NFTs in NFT marketplaces, including OpenSea, Rarible, SuperRare, and Nifty Gateway.

Chapter 3 of *Blockchain or Die* covered the importance of paper trading. To recap, whenever I invest in a new type of investment, I always paper-trade for at least one workweek to better understand the pricing and market share. Although I have been investing in cryptocurrencies since 2017, I considered NFTs a relatively new investment category, so I started paper trading them.

In October 2022, there were 591 NFTs on CoinMarketCap. I did not know which one to paper trade. So, I chose five NFTs directly between 1 and 59 and ended up paper trading NFTs 295-300. The prices for the five NFTs are as follows:

NFT	10/12/22	10/13/22	10/14/22	10/15/22	10/17/22
TDROP	$.005151	$.0050	$.005181	$.005147	$.005204
POLC	$.02828	$.02774	$.02755	$.02747	$.03081
MEFA	$.0003079	$.0002934	$.0002891	$.0002891	$.0002669
KOL	$.00357	$.003724	$.003393	$.0033399	$.003418
MNG	$.00405	$.004818	$.003859	$.004171	$.004228

In 2023 and 2024, I revisited these NFTs. Here are the 2023 and 2024 NFT tables:

NFT	09/12/23	09/13/23	09/14/23	09/15/23	09/16/23
TDROP	$.001234	$.001241	$.001247	$.001251	$.001259
POLC	$.004689	$.004633	$.004889	$.004969	$.004869
MEFA	$.00005202	$.00005207	$.00005206	$.00005153	$.00005245
KOL	$.001857	$.001843	$.001847	$.001841	$.001847
MNG	$.0002176	$.002168	$.002158	$.002201	$.002213

NFT	04/12/24	04/13/24	04/14/24	04/15/24	04/16/24
TDROP	$.001234	$.001241	$.001247	$.001251	$.001259
POLC	$.004689	$.004633	$.004889	$.004969	$.004869
MEFA	$.00005202	$.00005207	$.00005206	$.00005153	$.00005245
KOL	$.001857	$.001843	$.001847	$.001841	$.001847
MNG	$.0002176	$.002168	$.002158	$.002201	$.002213

These tables indicate that five of the NFTs experienced significant value loss from October 2022 to April 2024. One and a half years is a long time to paper trade, but if you started paper trading these five NFT tokens before you started investing, you could have saved a substantial amount of investment dollars!

The NFT investment checklist is similar to the cryptocurrency investment checklist.

	(NAME OF NFT)	Strongly Disagree -1	Disagree - 2	Agree - 3	Strongly Agree - 4	Neutral - 0
1.	You understand the company's business model.					
2.	The NFT adds value to business, industry, or society.					
3.	You were able to read a well-written white paper.					
4.	The company has been in existence for at least one year.					
5.	The company has an experienced CEO and executive team.					
6.	The company submitted all of the required regulatory filings.					
7.	The company has a well-designed professional website.					
8.	The company has an active and responsive social media presence.					
9.	The company has active smart contract activity on NFT exchanges.					
10.	At least three independent publications or organizations have published positive					

	(NAME OF NFT)	Strongly Disagree -1	Disagree - 2	Agree - 3	Strongly Agree - 4	Neutral - 0
	reviews or comments about the NFT.					
11.	The NFT has a functional attachment that adds additional value to the to it.					
12.	Any other factors you feel are important to add to the checklist?					

In addition to the above checklist, here are some important investment considerations.

1. BE CAREFUL WITH HYPED-UP PROJECTS

Excessive hype around an NFT or high-pressure tactics to invest your money quickly are huge red flags. NFT creators might promote their project through online articles, podcasts, and social media posts, but extensive marketing does not always reflect the NFT's actual value. Especially if the marketing promises a guaranteed return on investment or hypes a projected future price, beware — that's a red flag to steer clear of this NFT!

A while ago, I was approached for legal advice by an NFT company seeking to raise investment funds through a token sale. During every virtual meeting, they tried to persuade me to invest in their NFT. So, I considered the information in their pitch deck not only as an attorney but also as a potential investor. They set the initial sale price of their NFT token at $0.77, which seemed unreasonably high for a token not backed by a fiat currency, asset, or commodity and without a track record or market capitalization. Most tokens start off at a value under one cent until they can build market capitalization. In fact, many tokens never reach a cent in price, and this company was asking for $0.77 as a starting price. I saw numerous red flags.

Remember, pressure to invest immediately in a business or project is a red flag that it may be a scam or a poor investment.

I asked for the business justification for the $0.77 price, and they provided an equation based on the total supply and projected sales when the NFT started selling on an exchange. They also asserted the $0.77 price was "very low" and I should "snap it up" right now while I could. I informed them that their NFT was overpriced and that the value of their token was closer to $0.00077 than $0.77. Of course, my analysis ended our conversations.

This story highlights an unfortunate practice of creating subjective financials and predictive analysis to improperly justify initial token prices. As a potential investor, you should ensure that the NFT company can demonstrate sound financial backing and a verifiable market analysis for the token's price. This company never created a token and went out of business. If I invested in their NFT, I would have lost all my investment.

Ultimately, NFTs designed to offer a return on investment are even riskier than cryptocurrencies. Decentralized finance companies have spent years establishing their cryptocurrencies as alternatives to fiat currencies. Cryptocurrencies can be used to purchase a wide range of items and services. In contrast, NFTs currently lack widespread transactional or commercial applications. Therefore, be extra cautious about any company trying to sell NFT investments — the investment risk may outweigh the potential rewards.

2. HAVE AN NFT INVESTMENT ENTRY AND EXIT STRATEGY

Just like any investment, the best way to determine when to take NFT profits is to have an entry and exit strategy in place. Designing an entry and exit strategy before purchasing an NFT removes emotion from the investment, enabling you to make wiser investment decisions rather than falling prey to market whims, fear, or greed.

3. TAKE TRANSACTION FEES INTO ACCOUNT

Gas or transaction fees are necessary to verify transactions on a blockchain. They are used to incentivize miners to verify transactions on the blockchain. Gas and transaction fees directly impact your investment profits — the rate of price increase in your NFT must exceed the purchase price of the gas or transaction fees, or you'll end up losing money.

4. TAKE PROFITS WHEN YOU HAVE THEM

Many NFT investors buy and hold NFTs, hoping they will have the same high rate of return as Bitcoin or the BAYC. The reality is that the NFT market is extremely volatile. Buy-and-hold NFT investors may hold on to their NFTs too long, only to have the market move against them. When your NFT is profitable, consider taking the profits; otherwise, you may be incurring losses.

5. DIVERSIFY YOUR NFT PORTFOLIO

Instead of investing all your money in a single NFT, consider buying different types of NFTs from different creators or sellers. If one NFT collection decreases in price, you'll have a diverse selection of NFTs in your portfolio that may increase your gains or offset your losses.

After completing the NFT Investment Checklist and considering the additional investment strategies, you'll be better equipped to identify an NFT with more profitable traits. The next step before investing is to start paper trading to get used to trading NFTs.

HOW TO EVALUATE THE CREDIBILITY OF AN NFT

1. **Review the NFT's website and social media.**

Whether you want to purchase NFTs for investment purposes or personal reasons, it is crucial to buy from credible creators. Even for personal reasons, if the NFT company goes out of business, you could lose all access to your NFT. Start by reviewing the NFT creator's website—it is one of the best indicators of their professionalism and intent. If the website seems unprofessional or hastily made, it could be a rug-pull scam, where creators quickly collect their NFT profits and shut down the company. In that case, your NFTs will probably be inaccessible and worthless, and you will lose most, if not all, of your investment.

Social media presence is another great indicator. Check the creator's activity and presence on platforms like 𝕏 and Discord. Are they posting actively on social media? When was their last post? Is the creator engaging with their audience or answering community questions? What are the followers saying? A lack of social media activity or a weak follower base can signal poor marketing efforts, insufficient resources to invest in marketing, and a corresponding decline in customer interest. If the NFT is not legitimate, the community will often sniff out the scam. The creator's answers to community questions will reveal how much they have invested in their social media presence.

Beware of scams that imitate existing NFTs, so purchasers will buy the fake and worthless NFTs. Scammers frequently use social media to create fake NFTs and counterfeit wallets, targeting NFT buyers and their followers in fraudulent schemes. For example, scammers have used NFT creators' Discord channels to scam their followers. It is crucial to manage your NFT social media presence effectively to prevent scammers from defrauding you with counterfeit NFTs.

Due to these scams, some NFT creators have shifted their focus away from platforms like Discord, opting instead to use their own websites to build their NFT communities. This approach is perfectly fine, provided the website is up-to-date, responsive to user comments, and fosters a sense of community.

2. **Look for an NFT Roadmap**

A roadmap is a key indicator of an NFT project's potential longevity and success. Smart NFT investors rely on roadmaps to determine whether the NFT is likely to appreciate in value after the initial reveal or sales. Without a roadmap, the NFT risks discouraging educated and experienced investors, as the lack of direction indicates lesser long-term viability or success. It is essential to note that roadmaps may vary, as there are numerous types of NFTs.

NFT LEGAL ISSUES

The final section of this chapter will examine the legal challenges associated with NFTs. The creation, purchase, and sale of NFTs have raised a host of new and complex legal issues, with legislatures, regulators, and courts still learning how to address, enforce, and resolve these emerging issues. NFT creators, sellers, and buyers must understand the basic legal issues to make informed decisions regarding their NFT transactions and investments.

Here are three essential facts about NFT ownership that you need to know before buying NFTs:

1. NFT purchasers only own the tokens associated with the NFT.
2. NFT purchasers do not own the NFT media file or a copy of the media file that the NFT represents.

3. Unless specifically agreed in a smart contract, NFT purchasers do not own the copyright, trademark, or any intellectual property rights associated with the NFT.

Once you understand these facts, you will understand what you actually own if you purchase an NFT.

NFT INTELLECTUAL PROPERTY RIGHTS

I have been contacted several times by NFT creators seeking recommendations to successfully develop and sell their NFTs. After listening to their ideas, I asked them one simple question: Do they have the legal and intellectual property rights to sell their NFTs? The usual response, after a short pause, was "no." An increasing number of creators are driven by the hype around NFT profits but lack the knowledge of the legal and contractual processes required to establish an NFT company or properly protect their NFTs. This is exactly why NFT creators should consult with an attorney experienced in intellectual property, contract law, and blockchain technology. A thorough NFT legal analysis is a crucial component in the planning and execution of NFTs.

To fully grasp NFT intellectual property rights, it is necessary to examine NFT intellectual property issues from two perspectives: those of the NFT buyer and the NFT seller.

NFT BUYER

So, you've just purchased your first NFT. Congratulations! The next step is to ask yourself why you bought the NFT. Did you buy it for your art, music, or gaming collection? Or do you plan to sell the NFT for a profit? This is a very important question for one simple reason: you may unknowingly violate intellectual property rights if you try to sell the NFT.

If you have already purchased and sold the NFT, you may wonder how this is accurate; after all, no one accused you of intellectual property infringement when you sold your NFT. There are two ways to answer this question: 1. The creator of the NFT is not interested in the legalities of your resale. They are more interested in making a profit. 2. The NFT marketplace is not enforcing IP protections.

Here are the steps I use to verify NFT intellectual property rights:

- Examine the NFT artwork or content. Look for any signs of copyright infringement, including logos, trademarks, or copyrighted imagery.
- Use NFT marketplaces that have verification processes for creators and listings to ensure the legitimacy of your NFT purchase.
- Thoroughly review the intellectual property policies of the NFT marketplace where the NFT is listed. These policies often outline how intellectual property rights are handled on the platform.
- Search intellectual property databases for registered NFT trademarks or copyrights. Examples include the U.S. Patent and Trademark Office (USPTO) and the World Intellectual Property Organization (WIPO), as well as any other government intellectual property agency. If you suspect that an NFT infringes on intellectual property rights, report it to the NFT marketplace and relevant government agencies. This can help protect others from potential infringements. You are advised to consult with an attorney, especially for expensive NFT investments.
- Confirm the identity of the NFT creator or seller. Legitimate NFT creators will have an established online and social presence, with a portfolio of their work. Scammers may use pseudonyms or impersonate creators.
- After confirming the identity of the NFT creator, contact them to verify whether they are aware of the NFT sale and have authorized it.

- Ask the seller for documentation proving their rights to the content. This could include copyright certificates, licenses, or contracts that demonstrate the seller's legal ownership or permission to sell the NFT.
- If you are investing in an expensive NFT, seek legal advice from intellectual property attorneys who can guide you on whether the NFT infringes on any copyrights or trademarks. Verifying the authenticity and compliance of NFTs is essential, as both buyers and sellers are responsible for ensuring IP rights are respected.

FLAGGED NFTS

To reduce NFT scams and copyright infringement, NFT marketplaces began protecting users by flagging NFTs that may be associated with scams or illicit activities. Once flagged by the market, the NFT may be removed to safeguard users and maintain the integrity and reputation of the NFT marketplace.[228]

NFT REGULATORY COMPLIANCE

There are many open questions regarding the regulation of NFTs. Chapter 4 covered regulatory compliance and legal issues related to cryptocurrencies. This section will discuss the following NFT regulatory issues:

- Anti-Money Laundering
- Data Protection, Data Hosting, and Information Safekeeping
- Fractionalized NFTs

---

[228] Copeland, Tim. "What is a Flagged NFT." The Block, September 23, 2023. https://www.theblock.co/learn/251479/what-is-a-flagged-nft (Accessed September 3, 2025)

NFTS AND ANTI-MONEY LAUNDERING

The buying and selling of artwork are significant avenues for money laundering. The high-end art market is largely unregulated, highly private, and its prices are highly subjective. Moreover, artwork can be moved across borders more easily than large sums of cash. But what if it became even easier to launder money through buying and selling digital art? How do governments fight against NFT money laundering?

To safeguard against bad actors using NFTs to launder money, the Financial Crimes Enforcement Network (FinCEN) proposed new AML regulations requiring businesses that issue or exchange NFTs to register with FinCEN and comply with certain reporting requirements. Unfortunately, this regulatory requirement is not commonly known in the NFT space, so most NFT creators are not in compliance.

Other countries may also consider imposing similar regulations in the near future. As such, companies in the NFT space need to stay informed and prepared to comply with any new regulations.

DATA COLLECTION, DATA HOSTING, AND INFORMATION SAFEKEEPING

Generally, data protection requirements include measures to ensure that data is collected, stored, and used securely and responsibly. These may include encryption, access control, and data minimization. Data protection requirements vary depending on the jurisdiction and the type of data being collected. Depending on the jurisdiction, organizations may also be required to obtain consent for the collection and use of personal data.

"Right to be Forgotten" vs. Blockchain Immutability

Data protection laws, such as the GDPR (General Data Protection Regulation in the EU), grant individuals the "right to be forgotten" and the right to rectify or erase their personal data.[229] Many industry experts, including me, view this as a direct conflict with the immutable nature of blockchain technology, which ensures that data once recorded cannot be easily altered or deleted. For example, if NFTs contain personal data (e.g., identity-linked credentials), this immutability can pose a significant regulatory challenge.

Transparency and Consent

Companies handling personal data in the NFT space must obtain explicit consumer consent and be transparent about how their data is processed. This can be complex to manage within decentralized systems.

ANONYMITY VS. KYC/AML

Some individuals utilize decentralized blockchain transactions to maintain anonymity. Still, the increasing scrutiny of NFT platforms for anti-money laundering (AML) purposes means that "Know Your Customer" (KYC) and AML regulations are being applied. This requires platforms to collect and safeguard user identification data, raising further privacy concerns and undermining anonymity.

By using digital tokens to represent data, NFTs ensure that data is immutable and secure. As such, they are the perfect tool for data hosting and information safekeeping. With NFTs, users can securely store important documents, contracts, and other digital assets without worrying about them being misplaced or stolen. Additionally, NFTs enable tracking and verification of data authenticity, providing an extra layer of security for users.

---

229 *General Data Protection Regulation*, Regulation (EU) 2016/679 of 27 April 2016, art. 17, *Official Journal of the European Union* L 119, at 1 (May 4, 2016).

FRACTIONALIZED NFTS

Initially, the SEC did not consider NFTs securities because the initial versions could not be divided into smaller parts. When fractionalized NFTs were created, the door was left open for them to be considered securities.

Fractionalized NFTs (F-NFTs) are tokens that are divided into multiple parts and sold individually. They are usually for very expensive NFTs. F-NFTs are often used in digital art, collectibles, and gaming to create a more accessible and affordable market. F-NFTs also allow users to diversify their portfolios and access a wider range of assets.

Regulators are still navigating the introduction and enforcement of NFT regulations and policies. There are some general ideas about what might be regulated, such as ensuring that people aren't selling fake NFTs or misleading investors. However, many uncertainties still persist. For investors and small businesses that create or use NFTs, it's best to stay informed about new developments. Consult with an attorney to protect your investment, especially if you are investing a large amount of money in NFTs or committing resources as a business.

FILING TAXES FOR YOUR NFT PROFITS, OR LOSSES

I previously highlighted how the IRS is the most important government agency for individual cryptocurrency users and investors. The same holds true for NFTs.

The IRS has clearly stated that NFTs are considered digital assets, which may require reporting income from them as taxable income.[230] Even though

---

[230] Internal Revenue Service. "Digital Assets." https://www.irs.gov/businesses/small-businesses-self-employed/digital-assets#:~:text=You%20may%20have%20to%20report,from%20digital%20assets%20is%20taxable. (Accessed October 28, 2025)

the top of the 1040 forms does not ask you if you own NFTs, the IRS still considers them digital assets.

If you buy or sell NFTs, how should you report your NFT profits or losses on your tax returns? As discussed in Chapter 5, global regulators continue to debate how to tax cryptocurrencies, and in recent years, NFTs have become part of the tax conversation. The same questions apply: how will governments tax NFT profits?[231]

NFTS TAXED AS CAPITAL GAINS

Capital gains tax is the most common form of taxation for NFT profits. When you sell an NFT for more than you bought it for, the profit is probably considered a capital gain. The tax rate on capital gains varies by country and can be influenced by the duration of holding the NFT before it is sold (short-term vs. long-term). For example, in the US, short-term capital gains are typically taxed over a period of two years or less. They are generally taxed at higher rates than long-term capital gains. Similarly, capital losses are the most common form of NFT losses.

To calculate capital gains, you need to determine the cost basis of the NFT, which is typically the amount you paid for the NFT. The cost basis includes the purchase price, transaction fees, and any other costs associated with the NFT purchase. NFT transactions involving parties in different countries can raise additional tax concerns, including the potential for double taxation. If you're dealing with NFTs in a foreign currency, exchange rate fluctuations can impact your taxable gains or losses when converting the proceeds to your local currency.

231 Ibid.

NFTS TAXED AS ORDINARY INCOME

The frequency of your NFT transactions can also determine your tax treatment. If you frequently buy and sell NFTs, tax authorities may classify your activities as a business or trading, resulting in different tax rules.

NFT tax issues can be complicated. I recommend consulting a Certified Public Accountant or a tax preparer with experience in NFT tax issues. If you sell an NFT, the income from the sale may be taxed as ordinary income. The same applies if you receive an NFT as payment for services.

Finally, it is advisable to maintain records of all your NFT transactions, including purchase price, sales price, dates, transaction fees, and any related expenses.

NFT COURT CASES

Plaintiffs are starting to file lawsuits to resolve NFT legal disputes. Some iconic artists (Jay-Z and Quentin Tarantino) and powerful sports and entertainment organizations (Nike and the National Basketball Association (NBA)) are parties to lawsuits that are significantly shaping the NFT legal landscape.

NBA TOP SHOTS CASE

In Jeeun Friel v. Dapper Labs, Inc. and Roham Gharegozlou (Friel v. Dapper Labs), Friel sued Dapper Labs in a class action lawsuit. This lawsuit revolves around a single main question: Do NFTs qualify as securities under U.S. federal law?[232]

In 2020, Dapper Labs sold NBA Top Shots Moments NFTs, digital video clips from past NBA games, on the Flow Blockchain. Jeeun Friel sued

---

[232] *Jeeun Friel v. Dapper Labs, Inc.*, No. 1:21-cv-05048-JGK (S.D.N.Y. Aug. 31, 2021).

Dapper Labs, arguing that NBA Top Shot Moments, which are essentially digital collectibles representing basketball highlights, are unregistered securities. Friel argued that the value of Moments depends primarily on the efforts of Dapper Labs and relies on their promises for future development and potential integration with other platforms. In essence, Friel's argument relied on the fact that Dapper Labs has established an investment contract that satisfies the Howey test. Dapper Labs countered that Moments are collectibles, not securities, because they don't meet the Howey Test.

On February 22, 2023, the U.S. District Court for the Southern District of New York denied Dapper Labs' motion to dismiss. The court held that plaintiffs adequately alleged that Dapper Labs violated Sections 5 and 12(a)(1) of the Securities Act of 1933 by offering NBA Top Shot Moments NFTs without a registration statement.

Friel v. Dapper Labs is the first case to apply the Howey test to the sale of NFTs. The court stated that the facts presented a "close call" and noted that its decision was based on the specific facts before it.[233]

To resolve the class-action lawsuit, Dapper Labs agreed to a $4 million settlement, while maintaining that they did not admit to any wrongdoing or violations of securities laws. In addition to the monetary settlement, Dapper Labs committed to certain business changes, including further decentralizing the Flow blockchain, allowing for the trading of Moments on other marketplaces, and implementing employee training on securities law compliance.[234] The settlement resolves the dispute, allowing Dapper Labs to focus on its core business while addressing concerns raised about the nature of its digital collectibles.

---

233 Ibid.

234 Friel v. Dapper Labs, Inc., et al. No. 1:21-cv-05837, Dkt. No. 87 (S.D.N.Y. October 28, 2024)

It's crucial to understand that this is a private settlement of a civil lawsuit and does not establish a legal precedent. Courts are not bound by this agreement in future cases. Despite the lack of precedent, some legal observers believe that this settlement, along with previous SEC settlements involving NFTs (such as Impact Theory and Stoner Cats 2), offers emerging "rules of the road" for NFT issuers. These include factors the SEC and courts may consider when evaluating whether an NFT could be considered a security.

ROC-A-FELLA RECORDS, INC. V. DAMON DASH

In Roc-A-Fella Records Inc. v. Damon Dash, Damon Dash attempted to sell virtual ownership of Jay-Z's "Reasonable Doubt" album as an NFT. Roc-A-Fella Records was co-founded by Jay-Z, while Damon Dash was Jay-Z's former business partner.

The court ruled in favor of Roc-A-Fella Records by issuing a restraining order, which stopped Dash's attempt to sell off Jay-Z's 'Reasonable Doubt' album as an NFT. Dash's attempt to sell virtual ownership of the album's copyright signified a breach of fiduciary duty, as well as an act of conversion (selling without the owner's consent) and unjust enrichment. Ultimately, the case was settled for an undisclosed amount.[235]

MIRAMAX, LLC V. QUENTIN TARANTINO AND VISIONA ROMANTICA INC.

In Miramax, LLC v. Quentin Tarantino and Visiona Romantica Inc. (Miramax v. Tarantino), Miramax sued Tarantino for breach of contract, copyright infringement, trademark infringement, and unfair competition.[236] Riding the popularity of NFTs, in 2021, film director Quentin Tarantino

---

[235] Brittain, Blake. Jay-Z Label Settles Lawsuit Over 'Reasonable Doubt' NFT. Reuters.com, June 13, 2022. https://www.reuters.com/legal/litigation/jay-z-label-settles-lawsuit-over-reasonable-doubt-nft-2022-06-13/ (Accessed October 28, 2025)

[236] Queen, Jack. Tarantino, Miramax Settle Copyright Suit Over 'Pulp Fiction' NFTs. Reuters.com, September 9, 2025. https://www.reuters.com/legal/tarantino-miramax-settle-copyright-suit-over-pulp-fiction-nfts-2022-09-09/ (Accessed October 28, 2025)

created NFTs of his 1994 hit movie Pulp Fiction on OpenSea. The Pulp Fiction NFTs consisted of digital images of sections from the Pulp Fiction screenplay, along with the director's commentary.

Tarantino argued that he had the contractual right to sell the Pulp Fiction NFTs because he wrote the screenplay, directed, and produced Pulp Fiction. Miramax countered that they held all rights to Pulp Fiction, except for the rights granted to Tarantino under the contract, which did not include NFTs. The Miramax v. Tarantino case boils down to the contractual and intellectual property rights granted to Tarantino in Miramax's Pulp Fiction contract.[237]

Many blockchain legal professionals, including myself, eagerly awaited the court's verdict; however, Tarantino and Miramax settled the case. The ruling in this case would have provided great clarity on the ownership of NFT intellectual property.

U.S. V. ETHAN NGUYEN

Rug-pulls are nasty and devious schemes. As defined earlier in this chapter, an NFT "rug pull" occurs when investors believe they are investing in a cutting-edge idea at the intersection of art, technology, and finance. In reality, they are being scammed.

U.S. v. Ethan Nguyen et al. (U.S. v. Nguyen) is an NFT rug pull case. The Department of Justice (DOJ) alleged that the defendants promoted NFTs of "Frosties." After raising money from investors, the defendants transferred the NFT profits and shut down their operations, essentially stealing the money of NFT purchasers. The DOJ also alleged that they

[237] Ibid.

stopped the owners of Frosties from another planned NFT rug pull advertised as "Embers."[238]

Since 2023, there have been no updates on the U.S. v Nguyen case, which most likely means the defendants pleaded guilty or accepted a plea deal. Regardless of the outcome, the U.S v. Nguyen case shows that the DOJ is prosecuting NFT fraud cases.

NIKE INC. V. STOCKX

In Nike Inc. v. StockX, Nike alleges that StockX, an online reseller of goods including sneakers, sold NFTs that allegedly violated Nike's intellectual property rights. Nike's complaint states: "Without Nike's authorization or approval, StockX is 'minting' NFTs that prominently use Nike's trademarks, marketing those NFTs using Nike's goodwill, and selling those NFTs at heavily inflated prices to unsuspecting consumers who believe or are likely to believe that those 'investible digital assets' (as StockX calls them) are, in fact, authorized by Nike when they are not."[239]

StockX argues that its use of Nike trademarks falls under the legal concept of fair use, as the NFTs represent ownership of the physical Nike products. Furthermore, since StockX owns the actual Nike sneakers, it has the right to resell them and create NFTs referencing them, protected by the first-sale doctrine.

In a partial summary judgment issued in early March 2025, a New York district court ruled that StockX is liable for the sale of 37 pairs of counterfeit

---

[238] Press Release. Two Defendants Charged in Non-Fungible Token (NFT) Fraud and Money Laundering Scheme. Justice.gov, March 24, 2022. https://www.justice.gov/usao-sdny/pr/two-defendants-charged-non-fungible-token-nft-fraud-and-money-laundering-scheme-0 (Accessed October 28, 2025)

[239] McDowell, Maghan and Shoaib, Maliha. Nike Sues StockX Over NFTs. VogueBusiness.com, February 4, 2022. https://www.voguebusiness.com/technology/nike-sues-stockx-over-nfts (Accessed October 28, 2025)

Nike sneakers through its platform. This included four pairs sold to Nike investigators and 33 pairs sold to a third-party customer, who later discovered they were fake. This was a significant win for Nike on its counterfeiting claims.

Ultimately, in August 2025, Nike and StockX reached a confidential settlement, and the case was dismissed. The Nike v. StockX decision could have had a significant impact on NFTs. A ruling in favor of Nike would have established a precedent for stricter enforcement of trademark rights in the NFT space. Conversely, a win for StockX would have established broader legal grounds for using trademarks with NFTs to represent ownership of physical goods.

HERMES INTERNATIONAL AND HERMES OF PARIS V. MASON ROTHCHILD

Hermes International and Hermes of Paris (Hermes) sued Mason Rothchild in the Southern District of New York (SDNY) for trademark infringement (Hermes v. Rothchild). Hermes alleged that Rothchild used the trademarked Hermes logo on their Birkin handbags on his MetaBirkins NFTs and sold them online. Rothchild argued that the two-dimensional digital tokens were a commentary on fashion's fur-free initiative and an act of artistic expression that is protected under the First Amendment. The SDNY court did not find Rothchild's First Amendment argument persuasive and ruled for Hermes, finding that Rothchild infringed on Hermes' trademark.[240]

LESSON LEARNED 1: Creating an NFT and calling it "art" doesn't automatically shield it from trademark infringement claims if it commercially exploits a famous brand's goodwill and creates consumer

---

[240] McCormick, Erin. Jury Rules Artist NFTs of 'MetaBirkins' violate Hermes' Trademark Rights. The Guardian.com, Feburary 8, 2023. https://www.theguardian.com/fashion/2023/feb/08/hermes-metabirkins-trademark-court-case-mason-rothschild (Accessed October 28, 2025)

confusion. The commercial intent and potential for consumer confusion are critical factors.

LESSON LEARNED 2: Brand owners have a strong basis for protecting their intellectual property. The traditional principles of intellectual property are adaptable to new technologies. Brands will actively police and enforce their marks in these emerging areas.

LESSON LEARNED 3: A simple disclaimer may not be enough to avoid liability if the seller's intent was to trade on a famous brand's goodwill. The court looked at the totality of the circumstances, including Rothschild's marketing efforts and internal communications, which suggested an intent to capitalize on the Birkin name.

LESSON LEARNED 4: Companies should proactively consider registering trademarks for digital goods and services that might fall within their "natural zone of expansion" or where consumer confusion could arise.

LESSON LEARNED 5: Highly famous and distinctive marks have well-resourced legal departments to make sure infringers do not dilute the brand's unique identity.

These court cases are starting to frame the legal issues surrounding NFTs. As the courts issue more rulings on NFT cases, we can expect increased changes and regulatory complexities in the NFT legal landscape.

SUMMARY

This chapter covered a great deal of information about NFTs, touching on their current developments, future potential, and regulatory issues. This chapter is designed to increase your probability of success in a high-risk NFT market. If you wish to purchase NFTs as an investment, the tools provided in this chapter will help you make an informed investment decision.

The ever-increasing popularity and proliferation of NFTs also indicate significant advancements in blockchain technology. They have undergone considerable evolution since gaining popularity a few years ago — the subsequent drop in prices has forced them to adapt. I believe the most important stage of NFTs is yet to come. Although still relatively new, NFTs are poised to redefine the future of art, music, gaming, and commerce. The next stage of NFTs is likely to play a significant long-term role in commerce.

By 2045, NFTs will have become part of daily life. They are no longer just digital collectibles but have evolved into a vast economic ecosystem.

The art world has been revolutionized. Galleries and museums now have virtual spaces, showcasing NFTs from renowned artists and emerging talents. Artists sell their work directly to collectors, eliminating intermediaries and ensuring they receive more profits with fewer expenses.

Musicians are releasing their albums as NFTs, allowing fans to own unique digital copies of their music, as well as other digital attachments that the musicians have included with the NFT. These NFTs can be traded, resold, or even customized by the owner.

Beyond art and music, NFTs are being utilized in a wide range of industries. Gaming companies are creating unique in-game items that can be bought, sold, and traded on NFT marketplaces. Fashion brands are releasing digital clothing lines that can be worn in virtual worlds. NFTs are also used in real estate to transfer and record deeds or titles, and at closings to facilitate the buying and selling of houses.

SUCCESS STRATEGY ACTION ITEMS

1. Visit the NFT marketplaces mentioned in this chapter, including: OpenSea, Rarible, Nifty Gateway, Super Rare, and Mintable.
2. Stay current on NFT updates, especially on X and Discord.

3. Attend NFT conferences, in-person or virtual.
4. Use the NFT Investment Checklist to decide which NFTs meet your investment criteria.
5. Paper trade the NFTs you are interested in buying for at least a week.
6. Research the NTF creator/company. As covered in this chapter, this includes reviewing the NFT's social media and websites.

CHAPTER 8
IS THE METAVERSE THE *NEXT FRONTIER* FOR EVERYTHING?

"The metaverse is here, and it's not only transforming how we see the world but how we participate in it – from the factory floor to the meeting room."

Satya Nadella
CEO, Microsoft

CHAPTER OVERVIEW

The above words by Satya Nadella are a powerful recognition of the transformative potential of the metaverse. While many metaverse thought leaders have highlighted the metaverse's technological aspects, Nadella underscores its profound impact on every level of a company, from the factory floor to the boardroom.

Spoiler alert: the metaverse *IS* the next frontier to everything! As explored in this chapter, the metaverse is poised to revolutionize virtually every aspect of life, including the way we communicate, interact, and conduct business with one another. Over the past three decades, social media and the internet have completely transformed the way we think. The metaverse will likely do the same, becoming an integral part of daily life.

This chapter offers a comprehensive overview of the metaverse, examining its origins, definition, and portrayal in television and film. It will also explore the metaverse's societal influence, how corporations are engaging with it, practical use cases, and the legal challenges it presents.

MY METAVERSE JOURNEY

Back in 2019, I first heard of the metaverse, but with my schedule packed with blockchain trainings, speaking engagements, and book signings, I didn't take the time to dig deeper. It wasn't until the spring of 2021 that the metaverse truly caught my attention.

After delivering a speech on blockchain legal issues at a conference, I sold and signed a number of *Blockchain or Die* books and answered questions from attendees. As the crowd thinned, two gentlemen approached my table and started asking questions about the metaverse. One of them shared how he had purchased property on the metaverse in 2020, which had almost doubled in value. Intrigued, I asked where he had purchased the property. He replied, "Decentraland." I congratulated him on his successful investment, and we exchanged information. I realized it was time to start seriously researching the metaverse.

That night, after finishing all the networking events, I dove into everything I could find on the metaverse. I also researched Decentraland, Sandbox, Roblox, and other metaverse sites. That marked the beginning of my journey into the metaverse.

The next step in my metaverse journey was hands-on exploration — using a metaverse site, linking it to my wallet, and creating my first metaverse avatar. I chose Decentraland as my starting point, and using Decentraland's avatar options, I created an avatar that best reflected my inner and outer identity.

My first few experiences in Decentraland were underwhelming. It felt like playing a video game, where I just moved from room to room without much interaction. I looked for events and transported to their supposed locations in Decentraland, but no one was there. It was clear that the metaverse still faced significant hurdles to achieve mass user adoption.

Next, I purchased a set of virtual reality (VR) goggles. The VR goggles changed my entire experience! First, VR goggles are a more immersive and engaging way to experience the metaverse. Additionally, using VR goggles offers more interactive options, making the VR experience far more engaging.

Over the years, I have noticed a conspicuous increase in metaverse traffic, indicating that user adoption is on the rise. As I attended more metaverse events, I watched businesses use it for product launches, customer engagement, training, and hosting large conferences in immersive virtual spaces.

METAVERSE USE STATISTICS

The metaverse is currently in the early adoption phase and holds immense potential for the future. Allied Analytics analysis clearly illustrates the rapid growth and expanding reach of the metaverse.

- The global metaverse is projected to reach $1,237 billion in 2030.
- Metaverse is projected to reach 2.6 billion users by 2030.
- The American is the leader in metaverse innovation and investment.
- The Asia-Pacific region is projected to witness the fastest growth.[241]

These statistics clearly show the value of the metaverse, both as an emerging technology and as a potentially lucrative investment opportunity.

---

[241] Allied Analytics, LLP. Metaverse Market to Hit $1,237 Billion by 2030, Driven by VBR, AR, and Digital Innovation Boom. EINNewswire.com, November 6, 2025. https://www.einnews.com/pr_news/8648/5129/metaverse-market-to-hit-1-237-billion-by-2030-driven-by-vr-ar-digital-innovation-boom (Accessed December 30, 2025)

As metaverse technology becomes more accessible and affordable, these numbers are expected to increase.

THE DEFINITION OF THE METAVERSE

The term "the metaverse" is actually a misnomer, as it gives the impression that there is one large global metaverse. In reality, several metaverse platforms already exist, and many more are expected to be developed in the years to come.

Metaverse technology has created the next "gold rush" for big profits. There are numerous definitions of the metaverse. To obtain a fuller understanding of the metaverse, let's compare a few definitions.

> The metaverse is a hypothetical iteration of the internet as a single, universal, and immersive virtual world that is facilitated by the use of virtual reality (VR) and augmented reality (AR) headsets.[242]

> A persistent virtual environment that allows access to and interoperability of multiple individual virtual realities. [243]

> The metaverse refers to the convergence of physical and virtual space accessed through computers and enabled by immersive technologies such as virtual reality, augmented reality, and mixed reality. Described by proponents as the next iteration of the internet, this 3D virtual world is envisioned as a persistent, collective, shared space where digital facsimiles

---

[242] Metaverse. Wikipedia. (Accessed October 28,, 2025)

[243] Metaverse. Merriam-Webster. https://www.merriam-webster.com/dictionary/metaverse (Accessed September 5, 2025)

> of ourselves, or avatars, move freely from one experience to another, taking our identities and monetary assets with us.[244]

These definitions are informative and instructive, but they are not complete. The metaverse is so novel and complex that most people and companies are still navigating its boundaries, resulting in various definitions and interpretations. However, the following definition in the book *The Metaverse* by Matthew Ball most accurately includes all its components:

> A massively scaled and interoperable network of real-time rendered 3D virtual worlds that can be experienced synchronously and persistently by an unlimited number of assets with an individual sense of presence, and with continuity of data, such as identity, history, entitlements, objects, communications, and payments.[245]

On comparing the above definitions, "cryptocurrency" is the only word that appears to be missing in Matthew Ball's definition. However, the term "payments" can be considered a catch-all term that encompasses cryptocurrencies, crypto-tokens, and even fiat currencies. In the broadest sense, payments are the best term to describe the monetization of a metaverse.

WHO CREATED THE TERM "THE METAVERSE"?

The word "metaverse" was first introduced in the book Snow Crash (1992) by Neal Stevenson. The book presents Stevenson's creative vision of the future, which intertwines reality and virtual reality. One thing is certain: after reading Snow Crash, you will look at pizza delivery employees and couriers,

---

[244] Tucci, Linda and Moore, John. What is the Metaverse? An Explanation and In-Depth Guide. TechTarget.com, March 22, 2024. https://www.techtarget.com/whatis/feature/The-metaverse-explained-Everything-you-need-to-know (Accessed October 28, 2025)

[245] Matthew Ball, The Metaverse: And How It Will Revolutionize Everything (New York: W. W. Norton & Company, 2022)

spelled "kouriers," very differently! This is how Stevenson describes Hiro, the protagonist, using the metaverse:

> Hiro is in a computer-generated universe that his computer is drawing onto his goggles and pumping into his earphones. In the lingo, this imaginary place is known as the Metaverse. Hiro spends a lot of time in the Metaverse. Hiro is approaching the Street. It is the Broadway, the Champs Élysées of the Metaverse. It is the brilliantly lit boulevard that can be seen, miniaturized and reflected backward, through the lens of his goggles. It does not really exist. However, millions of people are currently walking up and down it.[246]

Stevenson's depiction of futuristic technology and human interface describes the metaverse as it may exist in the near future. *Snow Crash* even acknowledges that some properties in the metaverse have more value than others, a reality that is already true today. Depending on their location on the metaverse platform, some properties are more valuable than others.

However, none of the current metaverse platforms have the bandwidth to support thousands of people in one metaverse room. For large events, metaverse companies must set up separate rooms to manage the bandwidth.

DOES THE METAVERSE NEED BLOCKCHAIN TECHNOLOGY?

This is the trillion-dollar question, especially given the title of this book — *Blockchain or Die Harder*! Do metaverse platforms truly require blockchain technology to operate? Since the introduction of Bitcoin, the battle between decentralized and centralized blockchains has taken center stage in technological discussions. The truth is, very few blockchains — aside from those used in decentralized finance — are fully decentralized. Many

---

[246] Ibid.

companies label their blockchain as "decentralized," but in practice, most operate as hybrid or even centralized blockchains.

As discussed in Chapter 6, it takes an average of 10 minutes to verify a transaction on the Layer 1 Bitcoin blockchain. While newer DeFi blockchains, such as Litecoin and Stellar-Lumens, offer faster transaction speeds, they still rely on miners to process transactions. Given the sheer volume of transactions, miners on fully decentralized blockchains often struggle to keep up with the demand.

Interestingly, a metaverse platform does not need a decentralized blockchain to function. All of the core functions of a metaverse environment can run on centralized servers. However, one component of blockchain technology is very prominent in the metaverse: tokenomics.

"Tokenomics," a blend of "token" and "economics," refers to the use and valuation of tokens within a system. In the metaverse, tokenomics plays a crucial role in creating incentives for users to purchase tokens that represent virtual assets. Tokenomics also enables funding and facilitates transactions on metaverse platforms, making them an important component of the ecosystem.

CoinMarketCap.com currently tracks the price of 50 metaverse tokens. At the time of publication, the top five metaverse tokens by market capitalization follow:

#60 – Render (RENDER): A provider of decentralized GPU-based rendering solutions. On top of a decentralized GPU computing network, it provides a platform for artists and developers to build services and applications for the emerging digital economy.

#88 – Stacks (STX): A blockchain solution for smart contracts, enabling applications and NFTs secured by Bitcoin.

#103 – FLOKI (FLOKI): A community-driven token inspired by memes, with a focus on developing its metaverse ecosystem.

#158 – MultiverseX (EGLD): Formerly Elrond, this token powers a scalable, efficient blockchain network designed for the metaverse and Web3 applications.

#168 – Axie Infinity (AXS): Known for its play-to-earn model, Axie Infinity's token supports a virtual world where users can collect, battle, and trade digital pets (Axies).

While blockchain technology isn't strictly necessary for a functioning metaverse, its integration — particularly through tokenomics — provides unique opportunities for new and exciting applications and innovative financial instruments. Metaverse tokens are not just digital assets; they represent a fundamental shift in how we exchange value and interact within virtual worlds. As blockchain technology and metaverse platforms continue to evolve, their integration has the potential to revolutionize digital spaces.

POP CULTURE AND THE METAVERSE

Is the metaverse an example of life imitating art, or art imitating life? Perhaps it's a bit of both. The concept of the metaverse has existed in pop culture for decades, long before the term itself was coined. Futuristic science fiction (sci-fi) books, movies, and TV shows have envisioned variations of what we now recognize as the metaverse. Despite their differences, these depictions share one commonality: the user is interacting with a computer program that has created a virtual or alternate reality. Below are some examples of how the metaverse has been portrayed in pop culture.

TRON

Tron (1982) was the first major motion picture to feature a major metaverse storyline. In the film, Kevin Flynn, a computer engineer, discovers that Ed Dillinger is stealing his work. Attempting to hack into the Tron mainframe using an old-school arcade video game, Kevin is digitized and enters the virtual world. Inside the Tron mainframe, he must confront Ed's virtual counterpart, Sark. Kevin finds allies in Tron and Yori, and together, they fight against the oppressive control of the system. Kevin's fight includes winning numerous physical contests against other challengers. [247]

In *Tron,* contestants wear helmets with goggles and full-body tech gear — similar to the VR/AR goggles and haptic vests seen in today's metaverse technologies. Life imitating art? Perhaps.

Tron was a cultural phenomenon that transcended movies and entertainment, inspiring a period of renewed fascination with computers and sci-fi. I even recall friends in school who started learning computer programming because of *Tron*!

STAR TREK

The holodecks from *Star Trek: The Next Generation* (1987), *Deep Space Nine* (1993), and *Voyager* (1995) represent a highly advanced versions of the metaverse. In these Star Trek shows, users interact with fully immersive simulations or avatars. The computer transforms matter within the holodeck, creating a lifelike environment based on the user's instructions. One of the main differences between the holodeck and the metaverse in the holodeck, users do not need goggles or an avatar; they personally interact with the holodeck environment.

---

[247] *Tron*, directed by Steven Lisberger (1982; Burbank, CA: Walt Disney Pictures), Film.

In case it isn't clear yet — yes, I am a Star Trek fan. However, bias aside, the original *Star Trek* (1966–1969) was remarkably ahead of its time. When the *Star Trek* TV show was rebooted in *Star Trek: The Next Generation*, the holodeck was a fascinating addition to the Star Trek universe. While its feasibility within the laws of physics remains doubtful, the holodeck is essentially a futuristic version of the metaverse: a mutually immersive experience in a virtual world.

THE MATRIX

In *The Matrix* (1999), which is set in a dystopian future, humanity is unknowingly enslaved by machines, serving as batteries to power the Matrix. Within this computer-generated reality, protagonist Neo, guided by Captain Morpheus and Trinity, fights to liberate humanity. [248]

Instead of AR/VR goggles, Mr. Anderson/Neo, like all humans created by the Matrix, is grown by the Matrix machines, serving as a human battery for the Matrix. The *Matrix* is a vast-scale technology that combines computer programs and machines. Under Captain Morpheus' leadership, Neo and his partner/love interest Trinity fight the Matrix computer program, which is protected by Agents, to free humanity from serving as batteries for the Matrix. Users must be "jacked in" to the Matrix, with a long spike inserted into the base of the skull. Once inside, they experience a hyper-realistic simulation. During Neo's kung fu training in the Matrix, Morpheus asks a profound question: "Do you think that is air you're breathing now?"—a poignant reminder of their simulated reality.[249]

What I find fascinating about the *Matrix* version of the metaverse is the unsettling notion that most humans don't even realize they're living in a

---

[248] *The Matrix*, directed by Lana Wachowski and Lilly Wachowski (1999; Burbank, CA: Warner Bros., 1999), DVD.

[249] Ibid.

simulation, that their reality is entirely virtual. They think they are living their actual life, not a virtual life. It begs the question: could there be a future version of the metaverse so immersive that users will not be able to distinguish between the virtual and the real world?

READY PLAYER ONE

In *Ready Player One* (2018), set in 2045, most of the world escapes reality through OASIS, a metaverse-like computer program. OASIS, an acronym for Ontologically Anthropocentric Sensory Immersive Simulation, is owned by Gregarious Games, the centralized metaverse monopoly in the future. When James Halliday, the owner of Gregarious Games, dies, a pre-recorded message announces a contest granting ownership of OASIS to whoever finds the digital Easter egg. Wade Watts and his friends join the dangerous and exciting treasure hunt for the digital Easter egg in OASIS.[250]

Unlike its predecessors, *Ready Player One* presents a more grounded depiction of the metaverse. There are no futuristic robots or space settings; instead, the narrative focuses on a near-future society where people use VR technology to navigate their virtual lives—a plausible extension of our current trajectory.

After seeing *Tron* and *The Matrix*, I had become accustomed to seeing movies with some connection to the metaverse. However, *Ready Player One* was unique because, compared to the other films and TV shows mentioned in this chapter, it is the most realistic portrayal of the metaverse. *Ready Player One* does not utilize holographic lighting effects; instead, it is not set in space or a future where robots dominate the world. In fact, most of the uploading to OASIS took place in the back of a raggedy van that had been retrofitted

250 Ready Player One, directed by Steven Spielberg (2018; Burbank, CA: Warner Bros. Pictures), film.

with virtual technology equipment. That is what makes *Ready Player One* a bit more realistic.

The above TV shows and movies showed how their version of the metaverse changed their imaginary world. But how is the metaverse changing society?

HOW THE METAVERSE IS CHANGING SOCIETY

The internet and social media have already revolutionized how we interact as a society. The metaverse takes this a step further by creating an immersive virtual society with its own rules, reshaping social dynamics, and influencing cultural norms. Metaverse communities are already developing new businesses, shaping new forms of relationships, and even identities. Here's how the metaverse is impacting key aspects of society.

GLOBAL COMMUNICATIONS

Three decades ago, we could not have imagined sending messages in real-time or video-calling at the tip of our fingers. The metaverse further enhances global communications by offering immersive platforms for collaboration and connection. People from different parts of the world can now interact in virtual spaces, breaking down geographical barriers.

ENTERTAINMENT AND MEDIA

Metaverse platforms are redefining entertainment and media consumption. Entertainment in the metaverse is highly personalized and interactive. Users can attend virtual concerts, participate in immersive gaming experiences, and even watch movies and shows in virtual cinemas. These innovations redefine how we consume media, making it more engaging and immersive.

ECONOMIC SYSTEMS

Metaverse platforms are creating new business opportunities, economic systems, and job opportunities. Businesses are already building and selling virtual assets, designing virtual environments, providing virtual services, or utilizing virtual representations to sell real-world items, such as clothing or cars. The metaverse is also generating jobs, from virtual asset creators to event organizers.

EDUCATION AND TRAINING

The metaverse is revolutionizing education and information sharing by offering hands-on, immersive training and learning experiences. Students can attend virtual classes, learn a trade, engage in simulations, and participate in experiments. These tools will expand access to education and vocational training for individuals who face geographical or economic barriers. For example, I have already started designing courses on a metaverse platform.

The metaverse is still in the early stages of development. Its future will depend on how it evolves, the governing policies and regulations, the cost of using metaverse platforms, and how companies, individuals, and communities use metaverse platforms.

CORPORATIONS AND THE METAVERSE

High-functional metaverse platforms will largely be built by corporations. Operating a functional metaverse platform requires a substantial amount of storage space and bandwidth, necessitating significant financial and technological resources. Corporations will need to invest liberally in technology bandwidth to build metaverse platforms that can scale to users' expectations. After the metaverse platform has been built, it must be marketed to attract users who want to buy and sell goods and services, attend events, play games, or simply hang out. Corporations will need to hire marketing professionals and utilize social media influencers to encourage users to create accounts and engage with their metaverse platform.

Diversity, inclusion, and belonging are important considerations in the development and implementation of metaverse platforms, given their global user base. To create an inclusive and truly communal environment, corporations must address and resolve the following issues:

COLLABORATION WITH DIVERSE COMMUNITIES

Collaborating with diverse communities and organizations ensures that a range of voices and perspectives are heard and represented in the development and governance of metaverse platforms. Engaging with diverse communities and seeking their input can contribute to a more inclusive and profitable metaverse experience.

REPRESENTATION IN DESIGN

Corporations should create diverse metaverse products that represent a wide range of races, ethnicities, genders, sexual orientations, abilities, and cultures. This includes offering diverse avatar options and designing virtual environments that reflect the real-world diversity of its users. To accomplish this, metaverse companies should employ a diverse range of developers, coders, business managers, and marketing professionals to build inclusive metaverse events and platforms.

CONTINUOUS EVALUATION AND EDUCATION

Regularly evaluating the metaverse's diversity and inclusion efforts, as well as gathering user feedback, can help identify areas for improvement. Promoting education and awareness about diversity, inclusion, and belonging within metaverse platforms enables users to understand and appreciate different perspectives and cultures, fostering a more inclusive and welcoming environment. This can be achieved through community events, workshops, and resources that encourage dialogue and learning. Making necessary adjustments based on feedback and evolving best practices can contribute to an inclusive and diverse metaverse.

Overall, successful diversity, inclusion, and belonging in the metaverse mirrors the strategic planning required in traditional diversity and inclusion efforts, including feedback from targeted diversity communities, diverse metaverse platform designs, thoughtful implementation, and ongoing executive commitment.

METAVERSE BUSINESS USE CASES

The metaverse offers a vast array of groundbreaking business opportunities by creating a virtual space where people can interact, create, and transact in ways that mirror real-world activities.

The metaverse has endless business uses. For example, individuals can buy virtual land, develop it, and sell it on the metaverse platforms. Businesses can sell virtual goods, such as clothing, accessories, and collectibles, for avatars. Brands can place ads on billboards and virtual cities metaverse platforms. Artists and entertainment companies can host virtual concerts, plays, and events, reaching a global audience without the constraints of physical venues. There's no limit.

From immersive virtual experiences to decentralized economies and digital asset ownership, the metaverse is revolutionizing traditional business models and opening new frontiers for innovation across industries. The metaverse represents a new frontier where creativity and commerce merge, revolutionizing how businesses and consumers engage in the digital age.

FROM "FACEBOOK" TO "META"

Mark Zuckerberg, CEO of Meta, the parent company of Facebook, recognized the vast potential of the metaverse and, according to BusinessInsider.com, legally changed the name of "Facebook" to "Meta" on October 28, 2021.[251] In the same year, Zuckerberg invested $10 billion in designing Meta's version of the metaverse.[252] However, Zuckerberg also faced reality by publicly admitting that the metaverse project would lose "significant" amounts of money over the next three to five years. While investing billions of dollars into metaverse platforms without any reasonable expectation of returns for three to five years is only possible for a multi-billion-dollar company like Meta, Zuckerberg's actions clearly indicate how we are on the cusp of a metaverse revolution.

META AND MICROSOFT

Meta and Microsoft have entered a partnership to develop metaverse platforms for both work and play. Microsoft sees the potential to combine digital and physical worlds to create metaverse platforms in the following industries:

AUTOMOTIVE METAVERSE PLATFORMS

Microsoft is developing metaverse platforms for the automotive industry. These platforms, centered on mixed-reality headsets and software, could have a significant impact on the industry in several ways, as discussed below.

---

[251] Canales, Katie. Facebook is Changing its Name to Meta. BusinessInsider.com. October 28, 2021. https://www.businessinsider.com/facebook-new-name-meta-rebrand-metaverse-zuckerberg-apps-2021-10 (Accessed October 28, 2025)

[252] Culliford, Elizabeth and Balu, Nivedita. Facebook Invests Billions in Metaverse Efforts as Ad Business Slows. Reuters.com, October 26, 2021. https://www.reuters.com/technology/facebook-revenue-misses-estimates-apples-privacy-rules-bite-2021-10-25/ (Accessed October 28, 2025)

Design and Development: The metaverse has the potential to revolutionize vehicle design and development. Automotive designers could create immersive virtual environments to test vehicle concepts, ergonomics, and user experiences before building physical prototypes.

Sales and Marketing: Can you imagine test-driving a virtual car before buying the actual one? The metaverse makes this possible. Automotive companies can utilize the metaverse to showcase new models in interactive virtual showrooms, enabling customers to explore vehicles in detail from anywhere. Virtual test drives could provide realistic driving experiences, which may enhance customer experience.

Aftermarket and Services: The metaverse could offer a platform for selling accessories, upgrades, and services. Customers could customize their vehicles virtually and purchase items directly within the metaverse. Additionally, virtual workshops could provide tutorials and support for DIY maintenance or repairs.

Training and Education: Automotive manufacturers could use the metaverse to train employees on new technologies, safety procedures, and customer service. Dealerships could also use it to train sales staff on product features and benefits.

The full implications of Microsoft's metaverse on the automotive industry are yet to be studied, but it definitely has the potential to transform how cars are designed, sold, and serviced. Of course, this would be limited to new cars using the Meta/Microsoft platform. It could take years, if not a decade, for the metaverse to be widely used in the automotive industry.

TRANSPORTATION METAVERSE PLATFORMS

Microsoft's metaverse platform, Mesh, has the potential to revolutionize the transportation industry in several ways. For instance, Mesh could significantly enhance remote collaboration among designers, engineers, and technicians, streamlining the development process and improving overall efficiency. Additionally, Mesh could facilitate virtual training for transportation personnel, allowing employees to safely practice complex tasks in a simulated environment, reducing both risks and costs.

Another interesting application is the visualization and optimization of transportation networks. By creating digital twins of transportation infrastructure, planners could experiment with different scenarios and identify potential bottlenecks or inefficiencies. This would enable better decision-making, optimize operations, and reduce costs.

Currently, Mesh's use in the transportation industry is primarily focused on industrial applications rather than consumer solutions. However, once the metaverse is fully implemented in the transportation industry, I see potential consumer applications. For instance, Mesh could help users monitor real-time traffic for automotive and public transportation and suggest alternative routes for faster travel.

TRAVEL AND HOSPITALITY METAVERSE PLATFORMS

Can you imagine experiencing vacation highlights before you book? Microsoft Mesh also holds significant promise for the travel and hospitality industry. By creating immersive virtual spaces, Mesh could transform how travelers experience destinations and make decisions. For instance, potential travelers could take virtual tours of hotels, explore attractions, or interact with digital representations of destinations before booking. This would enhance the decision-making process and reduce the risk of booking a disappointing experience. Additionally, Mesh could enable remote work and virtual conferences, making it easier for business travelers to stay connected while on the go. This functionality would cater to the growing trend of combining business and leisure travel.

However, Mesh's success in the travel and hospitality industry will depend on several factors. These include the platform's user experience, the availability of high-quality virtual content, and the level of adoption by both travelers and businesses. If Mesh can overcome these challenges, it could revolutionize the way people plan and experience their travels.

NVIDIA

NVIDIA, a leading technology company, has been actively developing metaverse projects to create more immersive, realistic virtual experiences. By providing the necessary tools and infrastructure, the company is contributing to a future in which virtual and physical worlds converge.

One of NVIDIA's most prominent products is Omniverse, a platform designed to create and simulate virtual worlds. Omniverse enables the real-time rendering of complex 3D environments, allowing architects, designers, and engineers to collaborate on projects and visualize their designs in a virtual space.[253]

Another notable project is Isaac Sim, a simulation platform specifically for robotics. Isaac Sim enables developers to train and test robots in virtual environments before deploying them in the real world, thereby accelerating the development of autonomous robots and AI systems.[254]

EPIC GAMES' METAVERSE

EPIC Games, best known for the popular game Fortnite, has been actively pursuing metaverse projects. Their vision is to create a shared digital universe where players can interact, socialize, and engage in various activities.

---

[253] NVIDIA Omniverse Real-Time Collaboration Capabilities Are a Hit. Engineering.com, July 11, 2021. https://www.engineering.com/nvidia-omniverse-real-time-collaboration-capabilities-are-a-hit/ (Accessed November 2, 2025)

[254] Ibid.

One of EPIC's most prominent metaverse initiatives is Unreal Engine 5, a powerful game engine that equips developers with tools to create highly realistic, immersive virtual worlds. Unreal Engine 5 is being used to build standalone metaverse experiences and integrate metaverse elements into existing games. Fortnite itself has served as a testbed for many metaverse features.[255]

EPIC's metaverse plans are still evolving. Although I am not a gamer, I view the company's focus on technology, partnerships, and community engagement as an indicator of a promising future for its digital universe. In addition, EPIC Games' vast experience in the gaming industry is another huge advantage, as gaming technology has synergies with metaverse technology. As the metaverse landscape continues to evolve, EPIC Games is well positioned to play a significant role in shaping this new era of gaming and social interaction.

METAVERSE LEGAL ISSUES YOU NEED TO KNOW

As discussed earlier in this chapter, the metaverse is creating new and complex legal issues. While this section may seem tailored to attorneys or legal professionals, it is equally important for users and creators in the metaverse. Why? Understanding the legal landscape can help you mitigate investment risks and provide the tools to make informed decisions about your involvement in the metaverse.

GOVERNANCE AND REGULATION

As the metaverse evolves and becomes more integrated into daily life, governments and regulatory bodies worldwide will likely draft legislation and

---

[255] Witt, Toni. Epic Games and Unreal Engine Join the Metaverse. CloudWars.com, April 27, 2022. https://cloudwars.com/metaverse/epic-games-and-unreal-engine-join-the-metaverse/#:~:text=Implications%20for%20the%20Metaverse,then%20businesses%2C%20then%20everyone%20else. (Accessed October 28, 2025)

regulations to strike a balance between innovation and consumer protection. However, the pace and scope of these regulations can vary significantly from country to country. They may be influenced by technological advancements, industry developments, societal concerns, and major metaverse companies. Key areas where governments might regulate the metaverse include:

CRYPTOCURRENCIES, NFTS, AND VIRTUAL ASSETS

Metaverse-based cryptocurrencies, NFTs, and virtual assets are likely to attract regulatory scrutiny regarding their classification, taxation, consumer protection, and anti-money laundering measures. For instance, metaverse cryptocurrencies might be classified as securities or currencies, and regulators such as the SEC may require platforms offering tokens to register them. This also extends to compliance with financial regulations, including KYC and AML requirements.

USER RIGHTS AND PROTECTIONS

Regulations might be introduced to protect users' rights within the metaverse, including privacy, security, fair trading practices, and protection against fraudulent activities. For example, establishing secure and reliable digital identities in the metaverse might require regulations to ensure authenticity and prevent identity theft or misuse.

The regulatory landscape can change rapidly, and as the metaverse evolves, so will the need for adaptive legislation. Questions about who has the authority to regulate activities within the metaverse and resolve disputes will require a mix of private agreements, platform-specific rules, and new legal frameworks. Staying informed about current events and policy changes is crucial for navigating this ever-changing environment.

INTELLECTUAL PROPERTY

Like other virtual assets, especially NFTs, metaverse virtual assets may require intellectual property protections to prevent unauthorized use or sale of the virtual assets. Protecting metaverse intellectual property rights and addressing issues such as copyright infringement, licensing, and ownership can be complex, but they remain grounded in fundamental intellectual property law.

SUMMARY

This chapter highlights the next stage in the evolution of technology, showcasing how tokenization in the metaverse is expanding globally in the realm of commerce.

Like the internet, the metaverse is poised to evolve over the course of decades. Its full potential will hinge on technological advancements, widespread user adoption, regulations, business metaverse products, ethical practices, and the transformative impact it has on users' lives and the broader world.

SUCCESS STRATEGY ACTION ITEMS

The following Success Strategy Action Items outline how I began engaging with the metaverse. I purchased my first pair of metaverse goggles on sale on Amazon Prime Day, and it has been a fulfilling experience ever since. So, use these action items to learn, explore, and have fun!

1. **Start with Social Media**: Many popular social media platforms are developing their own metaverse platforms. Explore these features to experience what the metaverse is like.

- Decentraland: This is a decentralized metaverse built on the Ethereum blockchain. It enables users to purchase and sell virtual land, as well as create their own unique experiences.
- The Sandbox: Similar to Decentraland, The Sandbox is a decentralized metaverse where users can create and monetize their virtual worlds and games.
- Meta (formerly Facebook): Meta is the most prominent player in the metaverse space, with its flagship platform, Horizon Worlds. It focuses on creating a fully immersive virtual world where users can socialize, play games, and attend virtual events.
- Roblox: While primarily known as a gaming platform, Roblox has elements of a metaverse, allowing users to create their own virtual worlds and experiences.
- Fortnite: Epic Games' Fortnite has been experimenting with metaverse features, such as virtual concerts and collaborations with popular brands.

2. **Experiment with Decentralized Applications (DApps):** Explore DApps built on blockchain technology, as many of them offer metaverse-related features. This will help you understand how decentralized platforms work and interact with virtual environments. The DAppList.com is a good place to find a metaverse DApp that you may like. https://thedapplist.com/category/metaverse

3. **Try Virtual Reality (VR) Games**: VR games provide immersive experiences that can help you grasp the fundamentals of virtual environments and interactions. Even if you don't own a VR headset, many online platforms offer free or low-cost VR experiences.
    - VRChat: VRChat enables users to create and explore virtual worlds, socialize with others, and engage in multiplayer games.
    - Rec Room: This platform offers a variety of social VR experiences, including multiplayer games, hangouts, and user-created worlds.
    - AltSpaceVR: While primarily focused on virtual events and meetings, AltSpaceVR also offers VR games and activities.
    - Horizon Worlds: Allows users to create and explore virtual worlds, socialize with friends, and play games.
4. **Attend Metaverse Events**: Similar to cryptocurrency, blockchain, and NFT events, metaverse event coordinators also coordinate metaverse virtual events, conferences, or workshops. These events will provide valuable insights and networking opportunities.
5. **Join Online Communities**: Connect with other metaverse enthusiasts through online forums, social media groups, or Discord servers. Sharing experiences and learning from others can accelerate your understanding.

Remember, the metaverse is a constantly evolving space — don't be afraid to experiment and learn as you go. The more you explore, the more comfortable you'll become with navigating this new digital world.

Part III
INVESTING IN CRYPTOCURRENCIES

CHAPTER 9

SHOULD YOU INVEST IN CRYPTOCURRENCIES?

"People have made fortunes off Bitcoin; some have lost money. It is volatile, but people make money off volatility too." [256]

Sir. Richard Charles Branson

British Billionaire

Sir. Richard Branson's words explain both sides of the volatility of cryptocurrency investment. Many people are making millions with cryptocurrencies and blockchains, but with every investment and business endeavor, especially those involving new technology, many also lose money. There are winners and losers in business and investment, especially in volatile markets. I hope that reading *Blockchain or Die Harder* will improve your chances for investment success and reduce your risk for investment loss.

CHAPTER SUMMARY

This chapter is designed to help you answer the critical question: *"Should I invest in cryptocurrencies?"* By addressing 13 commonly asked questions, this chapter delves deeper into various aspects of cryptocurrency investment, including altcoins, mining, staking, meme coins, NFTs, scams, and more, and provides strategies to maximize the benefits of this knowledge. These

[256] Jamie, Redman. Richard Branson: Blockchain Is an 'Economic Revolution'. Bitcoin.com, October 3, 2016. https://news.bitcoin.com/richard-branson-blockchain-revolution/ (Accessed September 4, 2025)

insights will guide your decision-making process, equipping you with the expertise to navigate the complexities of cryptocurrency investments.

COMMONLY ASKED QUESTIONS

As someone deeply immersed in the cryptocurrency and blockchain industry—whether as a bestselling author, speaker, attorney, trainer, or consultant—I'm often asked for guidance on navigating this complex and rapidly evolving space. Over the years, I've been asked certain questions more frequently than others by readers and audiences alike. Below, I've compiled the most commonly asked questions and provided clear, actionable advice to help you make informed decisions about cryptocurrency investing.

1. Should I invest in cryptocurrencies?
2. Is it too late to invest in cryptocurrencies?
3. Which cryptocurrencies should I invest in?
4. Should I invest in Bitcoin or alternative cryptocurrencies (altcoins)?
5. How do I find the next "Dogecoin" before it takes off?
6. Should I invest in meme coins?
7. Should I stake my cryptocurrencies?
8. Is it too late to start mining cryptocurrencies?
9. What are the differences between Bitcoin and Ethereum in terms of investment potential?
10. What is DeFi (Decentralized Finance), and how can I participate in it?
11. How can I protect my cryptocurrency from hacks and scams?
12. What impact will central bank digital currencies (CBDCs) have on traditional cryptocurrencies?
13. How can I use cryptocurrencies for everyday transactions, and where are they accepted?

1. SHOULD I INVEST IN CRYPTOCURRENCIES? ANSWER: IT DEPENDS.

As stated throughout these pages, cryptocurrencies are often considered very risky investments. In fact, in *Blockchain or Die,* I told a story about a doctor friend who invested in Bitcoin in 2015, when the cryptocurrency was trading around $50 per coin. Even though my friend would have made hundreds of thousands of dollars by holding onto his Bitcoin, he could not withstand the large price swings. He sold his Bitcoin, and to this day, he has no regrets. So, the decision to invest is solely up to the individual. If you are a risk taker, you might thrive in the volatility of the cryptocurrency industry. My doctor friend found 'peace' in selling his Bitcoin and eliminating the risk of losing his investment.

Similarly, numerous factors must be considered when deciding whether to invest in cryptocurrencies. Some investment factors are personal, such as your risk tolerance, the amount of money you can afford to invest, and the amount you can afford to lose. Some investment factors are market-specific, such as the current state of the cryptocurrency market. Is the cryptocurrency market in a bull market, a bear market, or in a sideways trading market?

Finally, you also need to consider certain cryptocurrency-specific factors. What is the market capitalization of the cryptocurrency? Is the cryptocurrency an established one with a long-term trading record, or is it new and unproven, with some consequential risk? The next story conveys a valuable lesson about personal risk factors.

CRYPTOCURRENCY INVESTMENT IN RETIREMENT

In mid-2021, I received a call from a retired family member seeking guidance on investing in Bitcoin and cryptocurrencies. I asked him if he knew anything about cryptocurrencies. He replied, "Only what you post on your Blockchain or Die Facebook page." I then asked whether he had a stream of income besides his retirement savings, to which he replied, "No." I gently advised him that I would not recommend investing his retirement savings in cryptocurrencies.

I recommended that he read about cryptocurrencies and blockchains, take training, listen to podcasts, and watch videos before making an investment. I also advised him to paper trade so he could get used to the volatility of the cryptocurrency markets. Finally, I told him how he could earn free cryptocurrencies, which will be covered in a subsequent chapter.

This was not the answer my relative wanted, but it was one he needed to hear. He was only looking for quick tips to buy those cryptocurrencies. Thankfully, I successfully discouraged him from investing in cryptocurrencies, as the market crashed on November 21, 2021, mere months after our conversation. My relative would have started his cryptocurrency investment with a substantial loss from his retirement account.

As I consistently emphasize in every speech, presentation, and seminar: never commit to an investment that you do not fully comprehend. This is followed by another cardinal rule of investment: never invest any funds you cannot afford to lose. I consider retirement savings as funds a retiree cannot afford to lose — they cannot be recovered in the event of a market downturn, nor are they replenished by new income. It is not the most prudent investment decision to put your retirement savings in a volatile market, especially a market you do not understand.

LESSON LEARNED 1: Don't invest in cryptocurrencies based on social media posts. Take the time to understand the cryptocurrency market and, more importantly, the cryptocurrencies you want to invest in. Only then should you invest in cryptocurrencies.

I generally use the following resources for my cryptocurrency investment research and information:

- CoinMarketCap: A comprehensive platform that provides real-time market data, news, and analysis on various cryptocurrencies.
- Coinbase: A cryptocurrency exchange that also offers educational resources, including blog posts and videos, to educate investors.
- Binance: A major exchange with a wealth of information, including research reports and market insights.
- Investopedia: A well-known financial website that offers articles and guides on cryptocurrency investing.
- Crypto.com: A cryptocurrency exchange and wallet provider that also has a blog with educational content and market analysis.

Find the investment resources that work for your investment and learning style.

LESSON LEARNED 2: Understand your risk tolerance and determine how much money you should invest. You should not risk investing money you cannot afford to lose. If you invest discretionary money and encounter a loss, your credit, household necessities, and lifestyle will remain unchanged. However, if you invest money you cannot afford to lose, you risk facing a personal financial crisis.

2. IS IT TOO LATE TO INVEST IN CRYPTOCURRENCIES? ANSWER: OVERALL, NO.

While watching a football game in the MetLife Stadium suite, a few attendees mentioned I was an award-winning blockchain author, speaker, and trainer, which piqued everyone's curiosity. One attendee approached me and stated that he'd never invest in cryptocurrencies because he believed it was too late — he had already "missed the boat." In response, I asked him if he invested in stocks. He did. I then asked him when he had last purchased any stocks. He replied that he'd purchased a few stocks a few weeks ago. I pointed out that the first stock ever traded was the Dutch East India Company in

Amsterdam in 1611, and the New York Stock Exchange was founded in 1817.

"If it's not too late to buy stocks more than 400 years after they first appeared," I said, "why would it be too late to invest in Bitcoin, which is only 17 years old, or other cryptocurrency companies that are not as old as Bitcoin?"

That perspective seemed to reframe his thinking about cryptocurrency investment.

But let's analyze this 13-year window from the perspective of two well-known publicly traded companies: Amazon and Netflix.

AMAZON (AMZN)

Should you have passed on investing if you missed the first 13 years of trading on Amazon? The following chart will help clarify that decision.

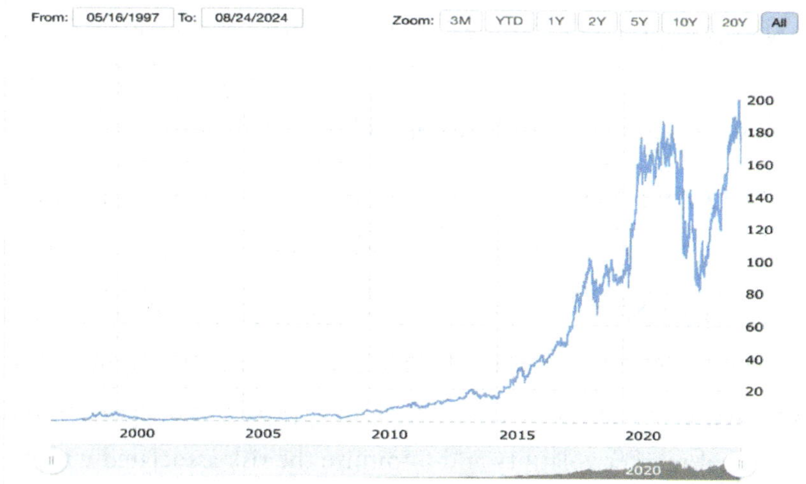

Amazon started trading in 1997 at seven (7) cents per share. Thirteen years later, Amazon traded at $6-$7 per share. As of July 2025, it was trading around $227 per share. It doesn't matter if you missed the first 13 years of trading Amazon stock — you would miss a $220-per-share increase if you decided not to invest in those 13 years of trading Amazon stock.

NETFLIX (NFLX)

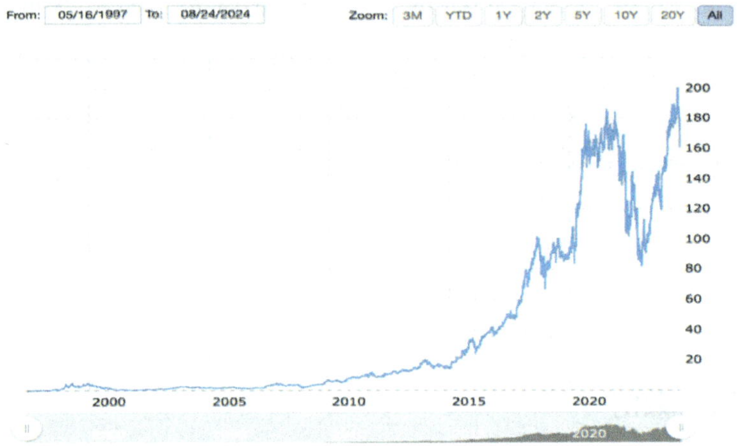

Netflix started trading in 2002 at $1.21 per share. Thirteen years later, in 2015, Netflix broke out, trading between $50 and $115 per share. As of July 2025, Netflix was trading at around $1,258 per share. You would have missed a $571 to $636 per share increase if you had decided not to invest, as you would have missed the first 13 years of trading Netflix stock.

LESSON LEARNED 1: It's never too late to invest in cryptocurrencies with a good business model and solid fundamentals at a good entry point.

LESSON LEARNED 2: Dollar-cost averaging (DCA) is an investment strategy where an investor regularly invests a fixed amount of money at set intervals, regardless of the asset's price. This approach is designed to reduce the impact of market volatility and minimize the risk associated with trying to time the market. Similar to the HODL strategy, DCA is particularly suitable for investors seeking a consistent and gradual investment approach.

Personally, DCA allows me to benefit from both sides of market trends. When prices rise, I enjoy the appreciation in value. When prices fall, I see it as an opportunity to acquire more of the asset at a lower cost, with the expectation of future price recovery.

I apply DCA to most of my investments, including cryptocurrencies. When I use DCA, I don't worry about perfectly timing my entry and exit points. As long as I feel the cryptocurrency will continue to increase in price, I will continue to use the DCA approach to investing in cryptocurrencies. DCA is also a good investment strategy for first-time cryptocurrency investors. With DCA, investors can purchase smaller amounts of cryptocurrency over time and buy on price dips, thereby adding more value to their cryptocurrency investments.

3. WHICH CRYPTOCURRENCIES SHOULD I PURCHASE FOR INVESTMENT? - ANSWER: IT DEPENDS ON MANY FACTORS.

Every investment has some level of risk, but some are riskier than others. Due to their price volatility and limited regulatory oversight, cryptocurrencies are considered high-risk investments. However, employing sound investment methodology to select your cryptocurrency investment can also mitigate risk factors and enhance your chances of investment success. Investment checklists are one method I use for all my investments.

Investment checklists serve as guardrails, preventing emotional decision-making and ensuring you consider crucial factors before purchasing stocks, bonds, cryptocurrencies, or other assets. They help you stay disciplined, analyze opportunities objectively, and build a well-rounded portfolio aligned with your risk tolerance and goals.

I always used investment checklists when investing in stocks. Still, because cryptocurrencies are a relatively new asset class, my stock market investment checklist needed several updates to remain effective. Here is the cryptocurrency investment checklist I designed to assess high-value cryptocurrency investments:

	(NAME OF CRYPTOCURRENCY)	Strongly Disagree -1	Disagree - 2	Agree - 3	Strongly Agree - 4	Neutral - 0
1.	You understand the company's business model.					
2.	The cryptocurrency adds value to business, industry, or society.					
3.	You were able to read a well-written white paper.					
4.	The company has been in existence for over five years.					
5.	The company has an experienced CEO and executive team.					
6.	The company submitted all of the required regulatory filings.					
7.	The company has a well-designed professional website.					
8.	The company has an active and responsive social media presence.					
9.	The company has a high market capitalization on many cryptocurrency exchanges.					
10.	At least three independent publications or organizations have published positive reviews or comments about the cryptocurrency.					
11.	The cryptocurrency is generating revenue or fees.					
12.	Are there any other factors you think are important to add to the checklist?					

A score of 48 is the maximum number a cryptocurrency can receive to determine whether it has personally met my investment standards. Overall, I will most likely not invest in a cryptocurrency with a score below 75% of the total.

In early 2021, at a party with several international friends, one of the attendees asked me if I was familiar with a European-based cryptocurrency. At the time, CoinMarketCap.com ranked this cryptocurrency around 1,000th by market capitalization, so I will refer to it as "Crypto1000."

He proceeded to tell me about Crypto1000, emphasizing its current price and offering a very optimistic prediction of where it would be. I told him I would look into Crypto1000 and get back to him.

A few weeks later, I ran into him again at another party. During the conversation, I shared my research, findings, and honest opinion about Crypto1000, including the fact that its business model did not make sense to me. He did not seem surprised, so I asked him if he knew the current price of Crypto1000. It turns out the price had dropped considerably, and he lost most of his money. Crypto1000 never returned to its initial purchase price, even though the market eventually rebounded.

My doubts about Crypto1000 stemmed from an analysis using my cryptocurrency investment checklist. Here are the results of my Crypto 1000 investment analysis:

	CRYPTO 1000	Strongly Disagree - 1	Disagree - 2	Agree - 3	Strongly Agree - 4	Neutral - 0
1.	You understand the company's business model.	X				
2.	The cryptocurrency adds value to business, industry, or society.	X				
3.	You were able to read a well-written white paper.	X				

	CRYPTO 1000	Strongly Disagree -1	Disagree - 2	Agree - 3	Strongly Agree - 4	Neutral - 0
4.	The company has been in existence for over five years.	X				
5.	The company has an experienced CEO and executive team.	X				
6.	The company submitted all of the required regulatory filings.	X				
7.	The company has a well-designed professional website.		X			
8.	The company has an active and responsive social media presence.	X				
9.	The company has a high market capitalization on many cryptocurrency exchanges.	X				
10.	At least three independent publications or organizations have published positive reviews or comments about the cryptocurrency.	X				
11.	The cryptocurrency is generating revenue or fees.	X				
12.	Any other factors you feel are important to add to the checklist?					X

Since Crypto 1000 received a 12 out of a possible 48 on my investment checklist, I did not view it as worthy of investment. If my friend had objectively analyzed Crypto 1000 or used an investment checklist, he might have saved his money.

This cryptocurrency investment checklist also applies to day trading. The primary difference between long-term investors and day traders is the amount of time they hold their investments. Day traders hold their investments for a much shorter period, typically days or even hours. Regardless of the time they hold their investment, day traders should conduct thorough research and have a solid understanding of the company before investing. Otherwise, they are gambling instead of making an informed investment decision. Alternatively, day traders can consider trading bots to provide trade instructions. However, they are sure to thoroughly research the trading bot company and track record before subscribing to the trading bot and using their investment algorithms.

While answering this FAQ, it is important to specifically address the eleventh point on the checklist: "The cryptocurrency is generating revenue or fees." During the early and mainstream phases of cryptocurrencies, which we covered in Chapter 1, these digital assets were relatively new. They lacked fundamental value or generated revenue for years, even a decade. "Fees" are important because they are usually the primary source of revenue for some cryptocurrency companies, and the more they collect in fees, the higher their value.

Cryptocurrencies are currently generating considerable revenue for investors. However, cryptocurrency fundamentals are not yet as readily accessible online as those of companies that issue stocks. TokenTeminal.com is an excellent resource for cryptocurrency investors to research the fundamentals of a cryptocurrency company, as it provides access to the company's financial statements on its website.

LESSON LEARNED: There are over 8,900 cryptocurrencies on CoinMarketCap alone. I cannot tell you which one you should purchase or which will give you the best returns. The best advice is to use the

Cryptocurrency Investment Checklist to determine whether you should invest in a specific cryptocurrency.

4. SHOULD I INVEST IN BITCOIN OR ALTERNATIVE CRYPTOCURRENCIES (ALTCOINS)? ANSWER: IT DEPENDS ON YOUR RISK TOLERANCE.

As a brief reminder, altcoins (or "alternate coins") are cryptocurrencies apart from Bitcoin. As stated above, there are over 8,900 altcoins listed on CoinMarketCap.com alone, so I am frequently asked this question. Most people ask this question for two reasons. First, they think they cannot afford to purchase Bitcoin, and second, they are looking for a cryptocurrency that can experience the same price increase Bitcoin has seen, from $0.003 to $117,000.00.

If you, like most people, can't afford to purchase one full Bitcoin at the current price of $89,933.26 (of course, at the time of writing this book), you can buy a part of a Bitcoin. A "Satoshi" (sat) is the smallest unit of Bitcoin that can be purchased. One Bitcoin equals 100 million sats. For one sat to be worth one cent, one Bitcoin would have to be worth $1,000,000.00.

Investing in altcoins may be riskier than investing in Bitcoin, especially when investing in a new altcoin with no established track record and a low market capitalization. The altcoin's business case is another important consideration. If the altcoin has a compelling business case, the investment in it might be worth the risk.

LESSON LEARNED: Your personal investment risk tolerance is the most crucial factor in deciding whether to invest in low-market-cap cryptocurrencies. If you are looking for substantial returns, which Bitcoin has already achieved, then consider investing in an altcoin. Thoroughly research the altcoin and utilize the Cryptocurrency Investment Checklist.

5. HOW DO I FIND THE NEXT "DOGECOIN" BEFORE IT TAKES OFF? ANSWER: DON'T TRY TO FIND THE LATEST HOT CRYPTOCURRENCY; FOCUS ON FINDING CRYPTOCURRENCIES WITH SOLID FUNDAMENTALS.

One of the questions I receive most frequently is about the next Dogecoin. Investors seek to identify the next hot cryptocurrency poised to deliver high returns. I never give investment advice, so my answer usually disappoints them: use the Cryptocurrency Investment Checklist.

Using the Cryptocurrency Investment Checklist will probably eliminate the vast majority of cryptocurrency investments, and for good reason. However, since three close friends asked me how they could find the next Dogecoin, I researched the possibility of a predictive model to identify cryptocurrencies with a greater chance of achieving exponential gains before the gains are realized.

My research indicated that Dogecoin's exponential gains were largely driven by social media posts from celebrities or influencers. You probably already guessed the influencer: Elon Musk, billionaire and President of Tesla.

Elon Musk has been known to tweet about cryptocurrencies, particularly Dogecoin, a meme-based digital currency. Musk's tweets and public statements had a significant impact on Dogecoin's price.[257]

In addition to his social media posts, he appeared on Saturday Night Live on May 9, 2021. During a "Weekend Update" segment, playing a financial expert named "Lloyd Ostertag," Musk was repeatedly asked by Weekend Update Anchors Michael Che and Colin Jost to explain Dogecoin. After

---

[257] Fallert, Nicole. Everything Elon Musk Has Said About Dogecoin. Newsweek.com, April 16, 2021. https://www.newsweek.com/everything-elon-musk-has-said-about-dogecoin-1584242 (Accessed October 28, 2025)

some attempts to describe it, Che finally asked, "So it's a hustle?" to which Musk replied with a laugh, "Yeah, it's a hustle."

While watching the Weekend Update segment, I immediately knew that Dogecoin's price was going to decrease. Minutes after the segment ended, my cryptocurrency and blockchain WhatsApp, Discord, and 𝕏 feeds exploded. Before the show, Dogecoin was trading at $0.72 per coin. After Musk's comments on Saturday Night Live, Dogecoin's price dropped by 29 percent in one hour.[258] Although Dogecoin was still a major cryptocurrency and ranked in the top 10 by market capitalization at the time of publication, it had not yet recovered to its all-time high of approximately $0.72 per coin.

LESSON LEARNED: Do not invest in a cryptocurrency only because of celebrity or influencer endorsements or social media posts — that's the wrong investment approach! If you see a celebrity or influencer posting positive comments about a cryptocurrency, your next steps should be to thoroughly research the cryptocurrency and use the Cryptocurrency Investment Checklist.

6. SHOULD I INVEST IN MEME COINS? ANSWER – NOT RECOMMENDED, BUT IT DEPENDS ON THE TYPE OF CRYPTOCURRENCY YOU WANT TO INVEST IN AND YOUR LEVEL OF RISK TOLERANCE.

Following the success of Dogecoin, I met many people who wanted to know if meme coins are a good investment. Meme coins are cryptocurrencies inspired by internet jokes and online culture. Doge and Shib Inu are two well-known meme coins.

---

[258] MacKenzie, Sigalos. Dogecoin Plunges Nearly 30% During Elon Musk's SNL Appearance. CNBC.com, May 8, 2021. https://www.cnbc.com/2021/05/08/dogecoin-price-plummets-as-elon-musk-hosts-saturday-night-live-.html (Accessed October 28, 2025)

Meme coins experience extreme price swings due to hype and social media trends. This is particularly true for meme coins that gain mainstream attention or celebrity endorsements. However, their value is often driven by hype and speculation. Many meme coins lack a solid business case, real-world applications, or business-related functionality, raising concerns about long-term sustainability and the potential for price crashes.

The devastating story of "I Like This Coin" (LIKE) is a striking example of meme coin hype. LIKE launched on March 17, 2024, with an estimated market cap of $577 million, and raised approximately $7.7 million during its pre-sale. However, the investors' dream meme coin quickly turned into a nightmare. In just eight hours, LIKE's value dropped by a shocking 90%, and the creators have not been heard from since then.[259]

According to an article by the CryptoPotato titled "A Shocking Number of Meme Coins Have Failed in 2024," there has been a 97% failure rate for meme coin projects since 2024. So, the question is: would you invest in an asset class with only a 3% success rate?[260]

Partly due to their reliance on hype and social media interest, meme coins are also more susceptible to pump-and-dump schemes. A pump occurs when a person or a group of 'whales' buys or convinces others to purchase large quantities of a cryptocurrency or an NFT to raise the price. When the price peaks, these individuals quickly sell off, making large profits and effectively

---

[259] S, Eddy. These Meme Coins Solana Fall After Only One Month. Cointribune.com, April 22, 2024. https://www.cointribune.com/en/crypto-these-meme-coins-solana-fall-after-only-1-month/#:~:text=Among%20the%20most%20striking%20examples,plummeted%20by%20over%2090%25! (Accessed October 28, 2025)

[260] Jones, Wayne. A Shocking Number of Meme Coins Have Failed in 2024: Study. Cryptopotato.com, August 18, 2024. https://cryptopotato.com/a-shocking-number-of-meme-coins-have-failed-in-2024-study/#:~:text=New%20research%20by%20Chainplay%20highlights,projects%20last%20about%20three%20years. (Accessed October 28, 2025)

lowering the cost. Investors who bought in at the 2021 peak price lost most, if not all, of their investment.

The following is a list of my "*Meme Coin Red Flags*" that investors should use to assess the meme coin. If I find any of the following information when researching a meme coin, I will refrain from investing.

- Many meme coins are created as jokes, without a business model or any real purpose beyond speculation. If a meme coin lacks a clear utility or a roadmap, it's likely a pump-and-dump scheme. Purchasing a meme coin because someone's face is on it is not worth the investment!
- Meme coins that rely solely on social media hype and influencer endorsements without providing any substantial information, such as a business case or a white paper about the purpose of the meme coin.
- Meme coins that promise "astronomical returns," "guaranteed returns," or claim to be "the next Dogecoin." Scammers often use these promises to lure in unsuspecting investors. If you see these or similar phrases in their marketing information, it is probably a scam.
- Meme coins with an anonymous development team. Huge red flag! Legitimate projects usually have transparent and identifiable developers and a verifiable track record. This will require you to conduct more in-depth research, but it is time well spent before making an investment.
- If the meme coin's liquidity isn't locked or the lock is set for a very short period, there's a high risk of a rug pull. Locked liquidity prevents developers from easily withdrawing funds from the liquidity pool.
- If the meme coin is only available on small, obscure exchanges, it may be a sign that it's not a serious project. Serious coins, even if relatively new to the market, utilize their funding and resources to secure listings on reputable exchanges.

Overall, meme coins may offer the potential for high returns, but they are very short-term, high-risk investments. Carefully research the meme coin, weigh the pros and cons, understand the underlying risks, and never invest more than you can afford to lose.

LESSON LEARNED 1: Meme coins are highly volatile and risky investments. Do not invest in meme coins solely based on celebrity or influencer endorsements. Thoroughly research the meme coin to decide if it is a solid investment.

LESSON LEARNED 2: If you want to purchase a meme coin, watch out for scams! Since meme coins rely heavily on social media, scammers use social media to lure meme coin investors. Do your research thoroughly, don't invest in new or obscure meme coins, and use different wallets for your cryptocurrencies and meme coins.

7. SHOULD I STAKE MY CRYPTOCURRENCIES? ANSWER: THAT DEPENDS ON YOUR DESIRE FOR LIQUIDITY.

Staking is the process of holding funds in a cryptocurrency wallet to support transaction processing on the blockchain network. Staking involves locking cryptocurrencies to receive rewards, typically in the form of interest payments in the same or additional cryptocurrencies. Staking enables users to participate in the blockchain network's consensus process, validating transactions and generating new blocks, thereby contributing to the network's security.

To receive the staking reward, investors or miners must purchase or own a minimum amount of cryptocurrencies, referred to as the "minimum staking requirement." You can choose to stake your cryptocurrencies and receive rewards proportional to the stake in the form of newly minted coins or transaction fees.

As an incentive to stake, reward interest payments are always much higher than bank interest rates; however, there are two major downsides. You must "lock away" your cryptocurrency while it is staked, which means you will not have access to it for a specified period. Also, since you cannot access your cryptocurrency, you cannot sell it during that fixed period. Why is this important?

Think about certificates of deposit (CDs) offered by traditional banks. With CDs, investors lock in their investments for a fixed period and receive a fixed rate of return. At the end of that period, the bank returns the initial CD investment, along with the accrued interest. Staking works the same way. Both are mid- to long-term investment strategies, which may be unsuitable for day traders or those with short-term investment objectives.

On the other hand, cryptocurrencies themselves are risky investments with a speculative rate of returns. If the cryptocurrency market crashes, you may not be able to sell your staked cryptocurrency while the price continues to decrease. If you aren't able to sell your cryptocurrencies in case of a market crash, are the rewards of staking still worth the risk? Consider these factors before you decide to stake your cryptocurrency.

LESSON LEARNED: Staking has become increasingly popular because it offers a fixed rate of return. However, remember that cryptocurrencies are high-risk investments with no guaranteed returns. If the cryptocurrency fails, the staking rewards end. In addition, the cryptocurrency company may not allow you to sell your staked cryptocurrency until the end of the staking lock, even if the price decreases. You should not stake your cryptocurrency if you plan to sell it in the near future.

8. IS IT TOO LATE TO START MINING CRYPTOCURRENCIES? ANSWER – NO, BUT IT ALSO DEPENDS ON WHICH CRYPTOCURRENCY YOU WANT TO MINE.

When Bitcoin was initially traded, all miners needed was a laptop or a desktop to mine Bitcoin. However, as Bitcoin's price rose, mining became increasingly competitive, and miners began using application-specific integrated circuits (ASICs) to improve their chances of successfully mining Bitcoin. ASICs, which are specifically designed for Bitcoin mining, are optimized to execute the SHA-256 hashing algorithm, a crucial component of Bitcoin's proof-of-work consensus mechanism. They are built to efficiently perform the cryptographic computations necessary to validate and secure transactions on the Bitcoin network.

As Bitcoin's price continued to increase, miners developed specialized mining rigs, and the final stage of mining evolution was the creation of large mining corporations with numerous rigs and a massive capacity to mine Bitcoin. These large corporations made it extremely difficult for individual miners to mine Bitcoin successfully.

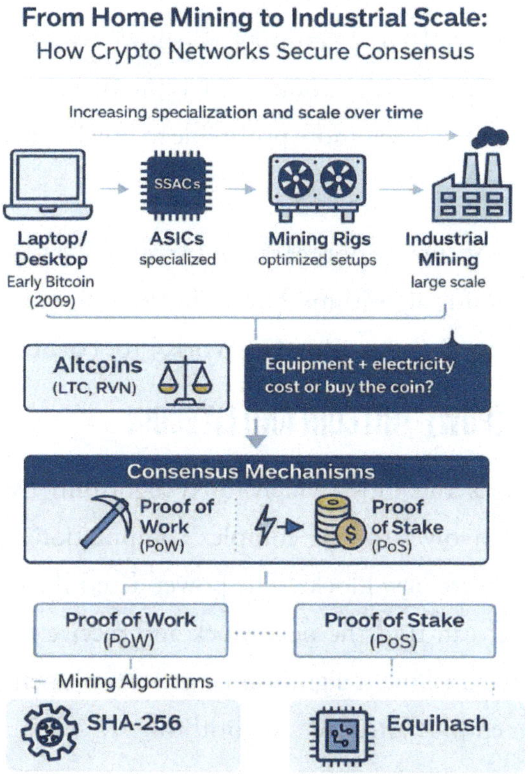

Successfully mining altcoins, such as Ethereum (ETH), Litecoin (LTC), Vertcoin (VTC), Ravencoin (RVN), or Aeternity (AE), is less competitive than mining Bitcoin. Individual miners have a greater chance of successfully mining these cryptocurrencies than Bitcoin. Before you start mining cryptocurrencies, consider this: Is it worth spending money to purchase mining equipment and paying the electricity bill to mine altcoins?

During the 2021 cryptocurrency crash, I strongly considered mining a top ten cryptocurrency on CoinMarketCap.com, let's call it Crypto10. I consulted with Esteban Mellid, a very experienced IT professional, for technical advice on mining Crypto10. We had a few meetings where we priced new ASICs with powerful graphics cards and estimated the electricity costs required to mine Crypto10. Finally, we factored in the cost of updating the graphics card every 18 to 24 months. When we compared our estimated cost of mining Crypto10 to the amount of Crypto10 we could purchase with that money, we realized it was more profitable to buy Crypto10 than to mine it.

The altcoin algorithm is another factor to consider when mining altcoins. There are many mining algorithms, but for illustration purposes, this section will cover the three major ones: Proof of Work, Proof of Stake, and Equihash.

PROOF OF WORK (POW) – BITCOIN AND LITECOIN

Currently, POW is the most widely used algorithm by cryptocurrency companies. PoW involves solving complex computational puzzles to verify transactions and create new blocks. The power determines how quickly and efficiently a miner can find the next block and receive rewards. However, POW computers consume a significant amount of energy, so many new cryptocurrencies employ alternative algorithms.

PROOF OF STAKE (POS) - ETHEREUM

In contrast to PoW, PoS algorithms reward miners based on the amount of cryptocurrency they hold. This reduces energy consumption but requires different hardware. For example, Bitcoin's Proof-of-Work consumes roughly 90 terawatt-hours annually — comparable to the energy use of a mid-sized country like the Netherlands. Ethereum, by contrast, reduced its energy use by over 99% after switching to Proof-of-Stake in 2022.

EQUIHASH - ZCASH AND MONERO

Equihash is designed to be ASIC-resistant, meaning it is more difficult to create specialized hardware (ASICs) for mining it. This helps maintain a more decentralized network, allowing individuals with diverse hardware setups to participate in the mining process.

DETERMINING COMPATIBILITY WITH YOUR MINING EQUIPMENT

To determine if your mining equipment is compatible with a specific mining algorithm, consider the following hardware requirements:

- CPU: Some algorithms, like Equihash, can be mined using CPUs.
- GPU: Many algorithms, especially PoW, are more efficient when mined using GPUs.
- ASIC: Specialized ASIC miners are usually the most efficient for PoW algorithms, but may be less flexible for other algorithms.

Once you know the altcoin's algorithm, you can coordinate the best setup to successfully mine it.

LESSON LEARNED 1: Individual Bitcoin miners face a great deal of competition from large mining corporations. Consider mining an altcoin to increase your chances of success.

LESSON LEARNED 2: Conduct your research before you start mining. Before purchasing mining equipment, it is essential to understand all the costs involved in mining cryptocurrencies. Understanding mining costs will provide more accurate profit estimates.

9. WHAT ARE THE DIFFERENCES BETWEEN BITCOIN AND ETHEREUM IN TERMS OF INVESTMENT POTENTIAL? ANSWER – BITCOIN AND ETHEREUM ARE THE TWO MOST PROMINENT CRYPTOCURRENCIES, BUT THEY OFFER DISTINCT INVESTMENT PROFILES.

Bitcoin is often referred to as "digital gold," primarily serving as a store of value. Its limited supply and decentralized nature have made it popular as a hedge against inflation and traditional financial systems. Although Bitcoin is a very volatile investment, it has significant long-term potential as a digital asset class.

Ethereum, on the other hand, is more than just a cryptocurrency. It is a digital business platform that enables decentralized applications (dApps) and decentralized autonomous organizations (DAOs) through the use of smart contracts. This versatility has led to its association with the burgeoning DeFi (Decentralized Finance) ecosystem. Ethereum's growth potential is closely tied to the success of decentralized applications (dApps) and the overall development of the blockchain industry.

Thus, Bitcoin's investment potential lies in its scarcity, growing adoption, and its role as a hedge against inflation, making it a relatively less risky long-term investment. Ethereum's investment potential, in contrast, lies in its business applications and versatility. Yet, it carries a relatively higher risk, as it relies on the widespread adoption of dApps and blockchain technology.

LESSON LEARNED 1: Your risk tolerance and investment approach should ultimately determine which cryptocurrencies match your investment strategy. If you prefer a more straightforward, store-of-value asset, Bitcoin may be the better choice. If you're excited about technology and business innovation and are willing to take higher risks for potentially high rewards, consider Ethereum.

LESSON LEARNED 2: Both Bitcoin and Ethereum can also serve as complementary investments in a diversified portfolio. Evaluate your investment goals and stay up to date on new developments in Bitcoin and Ethereum. Stay informed about Ethereum upgrades, such as ETH 2.0, and the growth of decentralized applications (dApps), as these factors will impact Ethereum's long-term value.

10. WHAT IS DEFI (DECENTRALIZED FINANCE), AND HOW CAN I PARTICIPATE IN IT? ANSWER – DEFI IS A FINANCIAL SYSTEM BUILT ON BLOCKCHAIN TECHNOLOGY THAT OPERATES WITHOUT INTERMEDIARIES, SUCH AS BANKS OR FINANCIAL INSTITUTIONS.

There are essentially four steps to participating in DeFi:

1. Understand Cryptocurrencies: Familiarize yourself with cryptocurrencies like Bitcoin and Ethereum, as they form the foundation of most DeFi platforms.
2. Choose a DeFi Platform: Research and select a reputable DeFi platform that aligns with your financial goals. Popular platforms include Uniswap, Aave, and Compound.
3. Create a Digital Wallet: You'll need a digital wallet to store your cryptocurrencies and interact with DeFi platforms.
4. Deposit Funds: Transfer your cryptocurrencies to the chosen DeFi platform.

DeFi offers a wide range of financial services, including lending, borrowing, trading, yield farming, and investing, all conducted directly between users. After setting up and funding your wallet, you can explore various DeFi services.

LESSONS LEARNED: The cardinal rule of investment applies to DeFi platforms as well: never invest in anything you do not understand. Conduct as much preliminary research as you can, and familiarize yourself with each of the four steps. If you feel uncomfortable with the first step, "understanding cryptocurrencies," don't proceed to the second step until you have a higher level of comfort.

11. HOW CAN I PROTECT MY CRYPTOCURRENCY FROM HACKS AND SCAMS? ANSWER – THERE ARE NUMEROUS TECHNICAL AND NON-TECHNICAL WAYS TO PROTECT YOUR CRYPTOCURRENCY.

Hacks and scams are one of the unfortunate dangers of investing in cryptocurrencies. No investment or financial instrument is entirely free from fraud or scams; however, cryptocurrency scams are distinct due to the technology involved. The following technical and non-technical measures will help you secure and protect your cryptocurrency investments.

1. Technical - Secure Your Wallet:

- Consider using hardware wallets. Hardware wallets, such as external drives and flash drives, are not connected to the internet and are considered the most secure option, as they store private keys offline.
- Choose reputable software wallets that offer robust security features.
- Always store your wallet's seed phrase in a safe, offline location. If you store your seed phrase on your computer, hackers might be able to access your seed phrase and then your cryptocurrency. Most importantly, never share your seed phrase with anyone!

2. Technical – Use Multiple Strong Passwords:

- Use unique, strong, and complex passwords that combine uppercase and lowercase letters, numbers, and symbols.
- Never use the same password for multiple accounts.
- Enable Two-Factor Authentication (2FA) whenever possible to add an extra layer of security. Although two-factor authentication (2FA) steps make accessing your account slightly more time-consuming, that extra step is a good way to secure your cryptocurrency.

3. Technical - Use Reputable Exchanges and Platforms:

- Select exchanges and platforms with a good reputation and strong security measures.
- Be cautious about exchanges that operate in jurisdictions with lax regulations.

4. Non-Technical - Be Careful of Phishing Attempts:

- Never reply to unsolicited emails, especially those asking for personal information or login credentials.
- Never click on links in emails or messages unless you are sure they are from a trusted source.

5. Non-Technical - Educate Yourself:

- Stay up to date on the latest security threats and best practices in the cryptocurrency space.
- Familiarize yourself with common scams, such as phishing, pump-and-dump schemes, and rug pulls.

6. Non-Technical - Diversify Your Investments:

- Avoid investing a significant portion of your portfolio in a single cryptocurrency or project.

- Conduct thorough research on any investment before committing your funds.

By following these guidelines, you can significantly reduce the risk of falling victim to cryptocurrency hacks and scams.

LESSON LEARNED 1: Scammers *are* trying to steal your cryptocurrency. The sooner you realize that, the quicker you'll make the right decisions to protect your cryptocurrency. Cryptocurrency scams should not be taken lightly, as they can result in substantial financial losses.

LESSON LEARNED 2: Protect your cryptocurrency using a combination of technical and non-technical measures. Using either one alone is not enough.

12. WHAT IMPACT WILL CENTRAL BANK DIGITAL CURRENCIES (CBDCS) HAVE ON TRADITIONAL CRYPTOCURRENCIES? ANSWER – CBDCS ARE EXPECTED TO HAVE A SIGNIFICANT IMPACT ON CONVENTIONAL CRYPTOCURRENCIES.

Central Bank Digital Currencies (CBDCs) are digital versions of a country's fiat currency, issued and backed by the central bank. Unlike cryptocurrencies, they are centralized and designed to be stable, replicating the benefits of physical cash in a digital world. Like all financial instruments, CBDCs have both pros and cons. While the exact nature and impact are uncertain, here are my perspectives on the pros and cons of CBDCs.

Pros:

- **Increased Payment Efficiency**: Transactions could become faster, cheaper, and more convenient, potentially reducing settlement times and operational costs. This can benefit individuals by lowering transaction fees, especially for cross-border payments and remittances.

- **Financial Inclusion**: CBDCs can provide access to digital payments and financial services for the "unbanked" or "underbanked" populations who may not have traditional bank accounts. This could be particularly impactful in emerging and lower-income countries.
- **International Transactions**: CBDCs could streamline cross-border payments, making remittances cheaper and faster for individuals sending money internationally.
- **Enhanced Security and Transparency (in some aspects)**: CBDCs can offer strong encryption and authentication protocols, potentially preventing fraud. The digital nature can also provide a clear audit trail, which could deter criminal activity.
- **Public Trust**: As they are backed by the central bank, CBDCs carry the same guarantees as a nation's paper currency, potentially fostering greater trust compared to cryptocurrencies. However, this strongly depends on the transparency the government maintains with its citizens and the trust citizens have in their government.
- **Resilience of Payment Systems:** CBDCs could provide a robust and resilient payment infrastructure, especially in extreme rural areas where banking is limited or non-existent, and in situations such as natural disasters or electrical grid issues, where traditional payment systems might be disrupted.

Cons:

- **Loss or Theft of Credentials**: As with any digital system, individuals could lose access to their CBDCs if they lose their digital wallet credentials (passwords, recovery phrases), potentially leading to permanent loss of funds.
- **Cybersecurity Vulnerabilities**: CBDC systems could be attractive targets for cyberattacks, hacking, and other security breaches, which could lead to financial losses for individuals and widespread disruption.

- **No Interest (Potentially)**: Many proposed CBDC designs do not offer interest or yield, meaning individuals holding CBDCs might not earn returns on their digital cash, unlike money held in traditional interest-bearing bank accounts.
- **Privacy Concerns**: Unlike cash, CBDC transactions can leave a "digital trail." Depending on the design, central banks or governments may have access to detailed transaction data, raising concerns about surveillance and potential abuse of power.
- **Potential for Account Freezing/Seizure:** Because the central bank directly issues CBDCs, theoretically, governments could freeze or seize individual accounts more easily and quickly than with traditional bank accounts, potentially for political reasons.
- **Programmable Money and Control**: CBDCs could be designed with "programmability," allowing for conditional payments, spending restrictions, or even expiration dates on money. While proponents argue this could be used for targeted policy functions (e.g., distributing stimulus funds), critics worry it could give governments unprecedented control over how individuals use their money.
- **Exclusion of the Digitally Underserved**: There's a risk that individuals without access to digital devices, reliable internet, or digital literacy could be further marginalized if CBDCs replace cash or become the dominant form of payment.

As you can see, CBDCs will probably present complications for any country that decides to implement them. The specific design choices made by central banks for their CBDCs will largely determine the ultimate benefits and risks for individuals.

IMPACT ON TRADITIONAL CURRENCIES

Now that we have covered the pros and cons, here are my thoughts on how CBDCs may impact traditional currencies.

As CBDCs are issued by the government, they are likely to avoid the same regulatory issues that cryptocurrencies face. In theory, CBDCs, backed by governments, offer stability and trust, potentially drawing users away from volatile cryptocurrencies. Additionally, CBDCs may also stimulate innovation in the cryptocurrency space.

CBDCs with privacy-preserving features could reduce the appeal of privacy-focused cryptocurrencies. Conversely, if CBDCs lack certain functionalities, such as programmability, cryptocurrencies could continue to thrive in specific niches.

Ultimately, the coexistence of CBDCs and cryptocurrencies is likely, with each serving distinct purposes. Cryptocurrencies may evolve into specialized assets for investment, payments, or governance, whereas CBDCs are used as a form of currency.

The specific lessons learned in this section will depend on how the country categorizes CBDCs. There are four potential country-specific lessons learned scenarios to consider when evaluating cryptocurrencies and CBDCs.

LESSON LEARNED 1: The government may allow cryptocurrencies but not issue a CBDC. In these cases, CBDCs will not exist to have an impact on cryptocurrencies.

LESSON LEARNED 2: The government may allow cryptocurrencies and issue a CBDC. This is the most complicated scenario, as both the cryptocurrency and the CBDC exist for use in the same country simultaneously. Government legislation and regulations will dictate how they are used as currency or as an investment.

LESSON LEARNED 3: The government may not allow cryptocurrencies and issue a CBDC. In these cases, the CBDC is the only digital currency that can be used.

LESSON LEARNED 4: The government may not allow cryptocurrencies and may not issue a CBDC. In this case, neither CBDCs nor cryptocurrencies exist to have an impact on the other, as neither exists in the country.

13. HOW CAN I USE CRYPTOCURRENCIES FOR EVERYDAY TRANSACTIONS, AND WHERE ARE THEY ACCEPTED? ANSWER – GENERALLY, CRYPTOCURRENCY ACCEPTANCE FALLS INTO TWO CATEGORIES: ONLINE RETAILERS AND BRICK-AND-MORTAR VENDORS. HOWEVER, IT DEPENDS ON INDIVIDUAL VENDORS.

While still in its early stages, cryptocurrency adoption is gradually growing, especially on online platforms. To use cryptocurrency for everyday transactions, you'll need a digital wallet to store and manage your cryptocurrency. Many online retailers, ranging from tech giants like Microsoft to clothing brands like Ralph Lauren, Banana Republic, and H&M, among others, accept popular cryptocurrencies such as Bitcoin, Ethereum, and Litecoin.

However, brick-and-mortar acceptance is more limited. Certain coffee shops and tech stores may accept cryptocurrency payments, but acceptance largely depends on individual vendors and stores. A workaround is to use cryptocurrency-linked debit cards, which convert your cryptocurrency to fiat currency for regular spending.

For example, Visa is collaborating with cryptocurrency exchanges to simplify the purchase of cryptocurrencies.

LESSON LEARNED 1: The acceptance of cryptocurrency can vary by location and over time. Before attempting to use crypto, always check a merchant's website or payment options. Additionally, platforms like BitPay offer a directory of businesses that accept cryptocurrency.

SUMMARY

This chapter is brief but important because it addresses the frequently asked questions I have been asked. These questions are also important because you may have considered these or similar questions.

To summarize the investment lessons and rules reflected through all of these common questions: Do your research! Understand the risks, never invest more than you can lose, and consider a diversified portfolio that may include some cryptocurrencies. Staking can be an option for some coins, but mining new ones might be less profitable now. Be proactive in securing your cryptocurrency and stay informed about the latest developments in the blockchain industry. Utilize both technical and non-technical methods to safeguard your cryptocurrency. Whether you're buying, selling, or using cryptocurrencies, a thorough analysis of each factor involved is key!

Even though each question in this chapter is concluded with a "Lessons Learned" section, here are some success strategy action items for you to complete.

SUCCESS STRATEGY ACTION ITEMS

1. Researching and obtaining knowledge in the cryptocurrency markets is very important. Read, watch, or listen to cryptocurrency resources for at least 10–15 minutes every day. This can be articles, blogs, v-blogs, social media posts, podcasts, and videos. Although some podcasts or videos may be 30 minutes to an hour or more in length, you'll find several shorter videos and podcasts as well. Many cryptocurrency resources are specialized for

specific industries, such as healthcare, supply chain, legal, identity management, government solutions, arts and entertainment, and many more.

2. Visit TokenTerminal.com to learn more about the revenue and fees generated by the cryptocurrency company. Visit the "Financial Statements" page to learn more about the fundamentals.

3. Use the Cryptocurrency Investment Checklist (CIC) on three or more cryptocurrencies to test if it works for your investment style. If you find that the checklist adds value, use it as a rule before investing in any cryptocurrencies. The CIC is only a model; feel free to modify the CIC to match your investment style.
4. If you are interested in mining, the NiceHash.com profitability calculator is a valuable resource for analyzing your mining profit potential. NiceHash.com is a global cryptocurrency hash power broker and cryptocurrency exchange with an open marketplace that connects sellers of hashing power with buyers utilizing a sharing economy approach.

5. If you have any spare change you want to convert into cryptocurrencies, consider using a Coinstar machine. You can find Coinstar machines in many supermarkets across the US. Alternatively, use the kiosk prompt on www.Coinstar.com to find a kiosk near you.

6. Bitcoin ATMs allow you to buy Bitcoin or other major cryptocurrencies without a cryptocurrency exchange. You can use Coinatmradar.com or similar sites to find a cryptocurrency ATM near you or anywhere in the world. https://coinatmradar.com

7. If you want to start using and purchasing cryptocurrencies through a debit card, the Bitpay Mastercard converts the cryptocurrencies into fiat currencies without a wallet or special apps. https://www.bitpay.com

CHAPTER 10

PROTECT YOURSELF FROM CRYPTOCURRENCY, NFT, AND METAVERSE SCAMS

"In the digital realm, where fortunes can be made and lost in a flash, the most valuable asset is not your crypto, NFT, or metaverse holdings, but your skepticism."

Eric Guthrie, Esquire

Author, *Blockchain or Die (2019) and Blockchain or Die Harder (2026)*

CHAPTER OVERVIEW

Scams involving cryptocurrencies, NFTs, and the metaverse have become increasingly common. While researching for this book, I surveyed nearly a hundred colleagues to see if they personally knew anyone who had fallen victim to these scams. Shockingly, all of them said "yes." Some even admitted, "Yes, and I was one of the victims."

This chapter examines the increasing prevalence of scams within these digital spaces, highlighting the tactics scammers use and offering practical strategies to safeguard your assets. Specifically, this chapter will:

- Examine why digital assets are attractive targets for scammers, including their rapid growth and potential for high returns.
- Discuss the most prevalent scams in the cryptocurrency, NFT, and metaverse worlds, such as phishing attacks, Ponzi schemes, and rug pulls.

- Provide a checklist of warning signs to look out for when investing in cryptocurrencies, NFTs, or metaverse projects.
- Offer practical advice on safeguarding your crypto wallets, NFT collections, and metaverse investments, including the importance of robust security measures and due diligence.
- Explain what to do if you believe you've been scammed, including steps for reporting the incident to authorities and exploring legal options.

The key takeaway is simple: don't just read this information. Apply these strategies to protect your cryptocurrencies, investment dollars, and personal information.

CRYPTOCURRENCY INVESTMENT SCAMS

A common theme in cryptocurrency investment scams is the false promise of high returns on investments, often accompanied by the victim's limited understanding and experience in the cryptocurrency market. Scammers claim they can quickly and easily provide huge returns for investors, but in reality, only they profit from the scheme. These scams are designed to steal your hard-earned cryptocurrency, personal information, or both. If the cryptocurrency investment seems too good to be true, it probably is.

CRYPTOCURRENCY SCAMS AND TECHNOLOGY

Look for the trap in three moves:

1. A slick site or app promising massive returns (100%, 200%, 300% or higher).
2. A small withdrawal that works, making you feel safe.
3. A sudden 'release fee' before you can cash out.

The moment you see step one, you're already in danger — stop before they reel you into steps two and three. These technology packages typically include websites, apps, an investment interface, social media presence, and often an email list or newsletter. The websites and apps have interfaces that allow investors to track their "profits," but these are fake numbers. Some investors even report making a successful small "test" withdrawal – just enough to convince them it was safe to invest larger sums of money. When investors try to cash out, they're instructed to send additional cryptocurrency to cover fake fees, or they receive no refund of their money.

Scammers also create fake websites that imitate real ones to steal your money. A good way to protect your cryptocurrency is to double-check the URL. Scammers often change a single letter or add another letter or number, relying on the fact that most people do not scrutinize the URL.

Two-Factor Authentication (2FA) has been addressed in earlier chapters, but the protection it provides warrants a mention here. Activate 2FA whenever possible to add an extra layer of security to your cryptocurrency wallet. While it requires additional steps to log in, this extra layer of security will be invaluable for protecting your accounts.

CRYPTOCURRENCY EXCHANGE SCAMS

Fake cryptocurrency exchanges are another prevalent scam. Investors transfer money to the exchange, thinking they are buying cryptocurrency, and even receive statements tracking the performance of their cryptocurrency. However, when they try to withdraw funds, they are asked to pay extra money or comply with false regulations before access is granted — if it ever is.

Resources like TechForing.com's article "The Ultimate List of Fake Crypto Exchanges: How to Stay Safe", highlight 122 allegedly fake crypto exchanges and scammers. Every cryptocurrency investor should use these resources before investing in cryptocurrency or sending money to a cryptocurrency exchange.[261]

261 (No Author) "The Ultimate List of Fake Crypto Exchanges: How to Stay Safe." Tech Foring, https://techforing.com/resources/articles/list-of-fake-crypto-exchanges/ (Accessed October 28, 2025).

CRYPTOCURRENCY RECOVERY SCAMS

Cryptocurrency recovery scams target victims who have already lost their funds to a previous scam. The recovery scammer promises to recover part or all of the stolen cryptocurrency, demanding upfront recovery fees. Ultimately, the cryptocurrency owner falls victim twice: first to the original scam, and then to the fraudulent recovery scheme.

There are many reputable cryptocurrency recovery services, but they usually focus on decrypting encrypted wallets, regaining access to your wallet, or recovering lost passwords. Legitimate cryptocurrency recovery services will only charge you if they actually complete the requested recovery service, and never ask for fees upfront.

The next section will cover cryptocurrency scams and the Federal Trade Commission.

FEDERAL TRADE COMMISSION

There are numerous scams in the cryptocurrency, NFT, and metaverse spaces; however, investment scams in the cryptocurrency space are the most prevalent. Since 2021, $575 million of all cryptocurrency fraud losses reported to the Federal Trade Commission (FTC) involved bogus investment opportunities. In reality, that number is much higher, as countless scam victims do not report being scammed to the FTC.

The FTC's primary goal is to protect consumers from deceptive practices and maintain a fair and competitive marketplace. Deceptive acts or practices include false or misleading statements or actions by businesses intended to deceive consumers. They can take various forms, such as false advertising, bait-and-switch tactics, pyramid schemes, and other forms of fraud. For instance, cryptocurrency companies that offer "guaranteed returns" would be a deceptive practice, as returns on investment can never be guaranteed or predicted with certainty in the cryptocurrency market.[262]

The FTC has been playing a very active role in monitoring, recording, and educating the public on cryptocurrency scams. Unfortunately, the data on cryptocurrency fraud is shocking. Since the start of 2021, more than 46,000 people have reported losing over $1 billion in cryptocurrency scams. These numbers only represent the scams that were reported to the authorities. Considering the number of people who prefer not to report scams, the actual number of cryptocurrency scams is much higher.[263]

The FTC data also shows that nearly 50 percent of cryptocurrency scams began with social media engagement, typically through an ad, post, or direct message. The top social media platforms identified in the FTC Report include Instagram (32 percent), Facebook (26 percent), WhatsApp (9 percent), and Telegram (7 percent).[264]

When you report a scam to the Federal Trade Commission (FTC), your information becomes a valuable tool in the fight against fraud. It is too late to recover your cryptocurrency or money, but you may be able to hold the bad actors accountable and prevent future scams.

Here's what typically happens when you report a scam to the FTC:

- Your report is added to the FTC's Consumer Sentinel database, a shared resource for law enforcement agencies worldwide.

---

262 Katte, Stephen. US Consumer Watchdog Probes Crypto Firms Over Deceptive Ads. Cointelegraph.com, December, 5, 2022. https://cointelegraph.com/news/us-consumer-watchdog-probes-crypto-firms-over-deceptive-ads (Accessed September 3, 2025)

263 Business Blog. Reported Crypto Scam Losses Since 2021 Top $1 Billion, Says FTC Data Spotlight.

264 Federal Trade Commission. "Consumer Protection: Data Spotlight, FTC Reporting Back to You." https://www.ftc.gov/news-events/data-visualizations/data-spotlight/2022/06/reports-show-scammers-cashing-crypto-craze (Accessed October 19, 2025)

- The FTC analyzes reports to identify trends and patterns in scams, enabling it to target specific fraudsters.
- If a large number of reports indicate a widespread scam, the FTC may launch an investigation and pursue legal action against the perpetrators.
- The FTC uses data to educate the public about emerging scams and how to protect themselves.

The next section will discuss the FTC's role in responding to cryptocurrency scams.

FTC SCAM ALERT ON DATING APPS AND CRYPTO SCAMS

Scammers also use dating websites and apps to scam victims. According to Statista, by the end of 2023, 441 million people worldwide used online dating platforms, making them a prime target for scammers.

In January 2022, the FTC issued an alert warning people about scammers who ask them to pay in cryptocurrency. The scam involves a romantic interest, a QR code, and a trip to a store (directed by the scammer while they are on the phone) to send money to them through a cryptocurrency ATM.

Picture this: you meet someone online or on a dating app, you like their photos, and you have a great conversation with them. You get to know them and might even become close over the course of a few weeks. They know you want to keep talking to them, so they ask you for money. The reason might be innocuous — they might need money for rent, medical bills, school, or even to visit you. You agree to pay them, and that's when cryptocurrencies enter the conversation. They call you and tell you to go to a store with a cryptocurrency ATM while they stay on the call with you. Once you're at the ATM, they direct you to insert your money into the ATM and buy cryptocurrency. They send you a QR code with their wallet address embedded. Once you purchase the cryptocurrency, they have you scan the code to transfer the money to them. They get the cryptocurrency, and you never hear from them again. They also ask victims to send money by directing them to a cryptocurrency machine to withdraw crypto and send it to their wallet.

Online dating scams may not directly impact you because you're married, in a relationship, or simply don't use a dating app. However, you probably know someone — a child, sibling, a friend, or a parent — who uses at least one online dating platform or app. This FTC Scam Alert provides dating app users with the information they need to protect themselves from online dating cryptocurrency scams.

Moreover, this scam is not limited to online dating. There are many kinds of scams in which bad actors try to steal your hard-earned cryptocurrencies. You may receive a call from someone claiming to be an official from a government agency, a local utility company, or law enforcement. After they get you on the phone, the same process follows, where they ask you to send them cryptocurrencies.[265]

HOW SCAMMERS TARGETED ME

Over the years, many scammers have attempted to hack, steal, and scam me out of my cryptocurrencies. I have received countless texts, calls, and spear-phishing emails from scammers posing as well-known companies and banks that I have never used, requesting payment on an invoice or asking me to verify my account information. Most people have received these phone calls or phishing emails. Below, I describe a few scams I have experienced in more

---

[265] Miranda, Christina. New Crypto Payment Scam Alert. Consumer.ftc.gov. https://consumer.ftc.gov/consumer-alerts/2022/01/new-crypto-payment-scam-alert (Accessed October 28, 2025)

detail to demonstrate the unexpected ways scammers can target you, and put you on your guard.

FEDEX OR UPS SPEAR PHISHING SCAM

The FedEx or UPS spear phishing scam is the most common method scammers have used to target me. At least once a month, I receive a spear-phishing email from FedEx or UPS, claiming I have a package waiting for me at a nearby distribution center, and asking for my personal information to verify my identity and collect the package. I receive these emails so often that if they were real, I'd have hundreds of FedEx and UPS packages that I never ordered waiting for me at numerous fake distribution centers. If I responded to their requests, scammers would use my personal information to gain access to my banking and cryptocurrency accounts.

CRYPTOCURRENCY REWARD SCAM

I also receive numerous direct messages from scammers saying I won varying amounts of cryptocurrency (the highest being $500). To collect the cryptocurrency, I simply need to download the app. Actually, the cryptocurrency is the bait, and the app is the hook. If I downloaded the app, the scammer would have access to all the information on my phone and could use it to access other wallets or exchanges.

EMAIL RAMSON SCAM

In 2022, one of my email accounts was hacked. The hackers sent me an email with my username and password as proof they hacked my account, and asked for $500 in Bitcoin, threatening to lock my email account if I did not comply. I never responded, and my email account was not locked. If I had, I'd probably have lost $500 in Bitcoin and some or all of my cryptocurrency.

Email ransom scams prey on victims' fears and vulnerabilities. Scammers typically employ a mass email campaign, claiming they can control and block recipients' accounts. The emails might also contain embarrassing photos or videos from their personal photo or video library, or compromising video footage of the recipient engaging in explicit online activities. To add credibility, they often include a password the victim may have used in the past.

The ransom email threatens to lock your email account or distribute the alleged photos or video to the victim's contacts unless a ransom is paid, usually in cryptocurrency. Email ransom scams are designed to evoke panic, making victims more likely to react impulsively and comply with the demands. It's crucial to remember that these emails are almost always fraudulent. Never pay a ransom — it does not guarantee the scammer will honor their end of the bargain. Instead, delete the email and report it to the appropriate authorities. Another way to protect yourself and your emails is to regularly back up your emails so that, in a worst-case scenario, you won't lose any of your emails or email attachments.

FAKE ACCOUNT UPDATING SCAM

In the spring of 2023, I received a call from a number that my caller ID identified as H&R Block. Although I'd never used H&R Block services, I answered the call. An automated message informed me that one of my cryptocurrency exchange accounts needed an update and asked me to verify my identity. In addition to the fact that I had never used H&R Block, it did not make sense for them to notify me that my cryptocurrency account had been hacked. I hung up right away, but if the scam fooled me, I would have given away all my account information and probably lost some or all of my cryptocurrency.

FAKE ACCOUNT RESTRICTED SCAM

In the fall of 2023, I received an email, seemingly from one of the cryptocurrency exchanges I use, notifying me that my account was restricted due to suspicious activity. The email appeared very genuine and even included a link to access my account. To be honest, it fooled me for a split second, but I had accessed that same account just a few hours ago to check my balances without any issues.

Additionally, I noticed a red flag.

The email stated: "Your account has been temporarily ***restrict.***"

Poor grammar, spelling, and punctuation errors are huge red flags that an email is not legitimate. If I had entered my account information, I would have given away my username and password, allowing the bad actors to steal all of my cryptocurrency.

FAKE VIRTUAL MEETING SCAM

One night in 2024, I received a call from a man representing himself as a Pastor with a prayer group and was invited to a Zoom prayer meeting. He got my name from one of the WhatsApp groups I'd joined years ago. He knew my name and some of the information from the group. I was immediately cautious and told him I could not attend the prayer meeting. He told me about upcoming prayer meetings and said he'd text me a link to join. However, his text was a WhatsApp code, not a Zoom or Google Meet invite. My WhatsApp notified me that he had sent me a message with a code, and he asked me to read it to him. The WhatsApp notification explicitly told me not to share this code with anyone. He kept urgently asking me to share the code, but I refused and ended the call. If I had given him the WhatsApp code, he would have had access to my phone and all its data.

Do any of these scams sound familiar to you? These scams are not limited to cryptocurrencies; scammers also accept gift cards as a form of payment. Both of these payment methods are outside the banking system and are difficult to trace, making them preferred payment methods for scammers.

JOB OFFER TEXT SCAMS

In 2024, I started receiving job offer scams, in which a person, usually a "woman" using her first and last name, claims that their company wants to hire me because they are in urgent need of employees. The job offer pays between $100 and $ 1,000 per day and includes full benefits and training. Some job offers even contain references to established job search websites, such as Monster and Glassdoor. All of this information is intended to set up the first part of the scam: using your personal information and accessing your money or cryptocurrency.

The text message will instruct you to respond via the same text message if you are interested in the position. If you respond, you have just verified that your cell phone number is active and in use. Once you respond, you will receive an e-form/application to fill out, which includes your name, address, and Social Security number. If you fill out the form, they will now have your personal information and can use it to access your identity and credit. If you let the scam get this far, you will probably receive a job offer. Here is where the second scam kicks in: you will be asked to pay for a laptop, phone, or training. This is a huge red flag. A reputable company will not ask you to pay before you join.

NFT SCAM EMAIL

In 2025, I received an email informing me that one of my NFTs had been successfully sold and that I should click the link in the email for more information. The problem is, I did not post my NFT for sale on OpenSea or any other NFT site when I received the scam email, so I knew it was a scam and did not open the link.

CRYPTO CALL CENTER SCAM

In 2025, one of the crypto exchanges I use sent a message stating that scammers had been impersonating its call center, requesting account information and personal details from users, and even pressuring them to transfer their funds. This scam takes it to the next level, beyond emails, texts, and apps. These call center scammers are trained to persuade you into providing them with your personal information or cryptocurrency. They exploit your fear of losing your crypto, and the irony is that they are the ones using your personal information to access your account and steal your crypto, or actually stealing your crypto in real-time by having you transfer it to one of their wallets.

STEPS TO AVOID CRYPTO COMMODITY SCAMS

This story shows how my research prevented a friend from buying into a scam.

A close friend asked for my advice about a cryptocurrency and commodity exchange he'd heard about from an acquaintance. Apparently, the exchange sent this acquaintance monthly cryptocurrency installments, and he was making a substantial amount of money. My friend wanted to invest immediately, but I convinced him to give me one day to research the company and provide feedback. Though my instinct said he was receiving staking rewards, I reserved my judgment until I completed my research.

1. I analyzed the cryptocurrency and commodity exchange's website and could not determine a clear business case or objective. That was my first red flag.
2. My second red flag was that I could not find a white paper.
3. Then, I went to SEC.gov, fully expecting the third red flag. Surprisingly, the company had received an SEC exemption, which gave me a glimmer

of hope. However, since the company brands itself as a "commodities company," I looked it up on nfa.futures.org and found nothing — third red flag!

4. Finally, I researched the company on social media, which provided valuable information. The CEO/Owner of the company had allegedly owned another cryptocurrency company that was investigated for fraud.

After only 10 minutes of research, I knew I would not invest a penny in this company, and I strongly advised my friend not to invest either. A year later, he thanked me for my advice — the acquaintance who'd told him to invest in the cryptocurrency company lost all of their investment. I have conducted variations of this research at least 100 times, for myself and clients, to identify cryptocurrency scams. As a result of this research, I have not invested in a single cryptocurrency scam.

This story is important not only because of the red flags but also because it demonstrates how anyone can utilize government websites and social media platforms to identify fraudulent cryptocurrency companies. If you use the same process, you will avoid cryptocurrency scams and investing in a fraudulent cryptocurrency or commodity company.

STEPS TO PROTECT YOUR CRYPTOCURRENCY

After receiving a barrage of suspicious emails every day, I have developed a three-step test to determine whether I should open any unsolicited emails, texts, or answer any calls.

1. **Do I know this person, or have I verified who I'm actually doing business with?** Like everyone, I am more comfortable opening emails from known sources. If I do not know the person or business that sent the email, I will not open the attachments or links in the email.

2. **Even if I do know the source, do they have a reason to email me?** If the email is unsolicited or does not make sense, such as a cryptocurrency exchange email claiming there is a problem with my account when I just viewed it an hour ago, I will not open the email, and I will definitely not click on any links.
3. **What do I stand to lose by opening the email?** Before opening an email from an unknown source, I must be 100 percent sure that I will not lose an opportunity by ignoring it. If I have any doubt about the source or information in the email, I do not open the email or any attachments. On rare occasions, when I open an email, I take the extra step to contact the business using the customer service number listed online, rather than the number provided in the email. Every time I have contacted the company to inquire about the urgent message they sent me, the representative has been unable to find any record of an email or a reason to contact me.

TECHNOLOGY CAN BE A SHIELD AGAINST SCAMS

Scammers can use technology to try to steal your cryptocurrency, but you can use technology to protect your cryptocurrency as well. There are also ways to utilize technology to safeguard your cryptocurrency against scammers. Here are some important recommendations.

1. **Use hardware/cold wallets:** There is a famous blockchain saying: "Not your keys, not your crypto." If you don't hold the private keys to a wallet, the centralized exchange can go out of business and disappear with your funds. It is best to use cold storage, especially for large amounts of cryptocurrency.
2. **Use multi-factor authentication:** Your wallet is as secure as your weakest link to it. Multi-factor authentication (MFA) strengthens your

wallet's security. The extra protection is worth the extra step and added time it takes to use MFA.

3. **Don't use your main wallet to interact with multiple dApps when trading**. Multiple secondary wallets allow you to engage in other cryptocurrency transactions without risking the funds in your main wallet. Your secondary wallets should contain minimal funds and be used to interact with dApps or smart contracts that you don't trust.

These technology-based recommendations will help protect your cryptocurrency. The last section of this chapter will cover NFT scams.

NFT SCAMS

Open a search engine, and type "free NFTs" in the search bar — you'll find hundreds of results with free NFT recommendations. Like airdrops, many "free" NFT websites and apps are scams that aim to collect your personal information. After you enter all your information on the airdrop website, you may not even receive an airdrop token — essentially, you have exposed your identity and personal information for nothing! This section will cover strategies to identify and protect yourself from NFT scams.

Here are a few ways NFT airdrop scams may operate:

Phishing Websites: Scammers set up fake websites that mimic the official NFT websites, asking users to provide sensitive information such as private keys, wallet addresses, or personal data to receive the airdrop.

Social Media Impersonation: Scammers create fake social media profiles, especially on platforms like X or Discord, pretending to represent reputed NFT projects. These accounts are then used to spread false information about airdrops and deceive users.

Malicious Links: Scammers share links to malicious websites or phishing pages through email, social media, forums, or direct messages.

Malware is downloaded onto devices when users click on links or prompts that trick them into providing sensitive information.

Token Swap Scams: Scammers may ask users to swap their existing tokens for tokens from a new NFT project to participate in an airdrop. Users who fall for this may lose their tokens without receiving the NFT airdrop.

Fake Wallets: Scammers promote fake wallet applications that claim to support the upcoming airdrop. Users who download and use these wallets may expose themselves to the risk of having their cryptocurrency stolen.

False Airdrop Announcements: Scammers create fake announcements with official-looking graphics, logos, and language to advertise a popular NFT airdrop. Once you give them your information or download their app, you have exposed yourself to the NFT scam.

To avoid falling prey to NFT airdrop scams, verify information from official sources. You can protect yourself from these scams by knowing which free NFT apps, websites, and airdrops are real and safe to use. Thoroughly research the NFT on social media. If the NFT is a scam, users will likely issue warnings on numerous social media sites.

Be careful with unsolicited messages or links, and never share sensitive or personal information or private keys with unknown entities. Additionally, remain informed about the latest scams — the creativity of scammers knows no bounds. Remember, if something sounds too good to be true, it probably is a scam.

METAVERSE SCAMS

As with all new technologies, the metaverse is rife with fraud, scams, and other forms of exploitation that can result in financial losses, identity theft, or other unfortunate consequences for users. A significant amount of communication, including warnings, to metaverse users and investors can help prevent hackers and scammers from finding more victims.

Metaverse scammers can use the same scams covered earlier, including:

- Cryptocurrency investment
- Airdrop
- Ponzi schemes
- Phishing and false websites
- Dating
- Rug pull
- Social media impersonation

Additionally, avatar impersonation is a new type of scam specific to the metaverse scams, a more specialized version of social media impersonation. Scammers clone your friend's avatar/name and DM you for 'urgent help.'

Rule: Never transfer assets in DMs. Confirm identity via a second channel (voice/video) and use only platform-approved trade or escrow services. Metaverse users should verify the identity of the person they are interacting with before sharing any sensitive information or completing a transaction.

Roblox Trading Scam

My friend's teenage son — let's call him Carson — has been very active on Roblox, a platform where you can build and trade virtual items, for years. Another Roblox user spotted Carson in the trading area and offered to trade one of his expensive items for everything Carson had spent years accumulating.

Carson agreed to the trade. However, before they could begin, the other user told Carson that his inventory was full. He requested a minute to remove some Roblox items and asked Carson to send his items over in the meantime. As soon as Carson sent his items over, the other user logged out. Carson never saw him or his Roblox items again.

While this scam did not result in financial loss or cryptocurrency theft, Carson was the victim of a Metaverse predator who took advantage of his trusting nature. I am sure there are thousands of similar stories.

SUMMARY

Imagine stepping into a digital wonderland where you can own unique, one-of-a-kind assets, create art or music, interact with friends in virtual worlds, and even make a fortune. Sounds like a dream, right? Unfortunately, this dream can quickly turn into a nightmare if you're not careful. I have many friends and associates who have been victims of cryptocurrency, NFT, and metaverse scams, underscoring the pressing need for greater awareness of scams in the digital assets space.

In this chapter, we've explored the evil side of the crypto, NFT, and metaverse worlds. From pump-and-dump schemes to rug pulls and phishing scams, there are numerous ways scammers attempt to steal your hard-earned money. As you navigate these spaces, remember that skepticism and knowledge are powerful tools to protect your investment and identity. By using technology to secure your accounts and transactions, conducting extensive research, understanding these common scams, and practicing safe online habits, you have the power to safeguard yourself and others from becoming the next victim. So, stay vigilant, use the information in this chapter, and enjoy the exciting world of crypto, NFTs, and the metaverse without falling prey to the traps set by the scammers.

SUCCESS STRATEGY ACTION ITEMS

The following action items will help you protect your personal information, money, and cryptocurrency.

1. **Conduct Due Diligence**
    - Research Projects: Thoroughly research any cryptocurrency or investment opportunity before investing.
    - Beware of Scams: Be wary of promises of guaranteed returns, high-pressure sales tactics, and unsolicited offers.
    - Verify Information: Always verify information from official sources and be skeptical of information from social media or online forums.
2. **Make Sure You Use Secure Connections**
    - Avoid Public Wi-Fi: Avoid accessing your cryptocurrency accounts on public Wi-Fi networks.
    - Use a VPN: If you must use public Wi-Fi, use a reputable VPN to encrypt your connection.
3. **Make sure you use strong passwords and 2FA.**
    - Unique Passwords: Use strong, unique passwords for each cryptocurrency exchange and wallet.
    - Two-Factor Authentication (2FA): Enable 2FA on all your accounts. This adds an extra layer of security by requiring a second form of verification (like a code from your phone) in addition to your password.
4. **Use Secure Storage**
    - Hardware Wallets: For long-term storage, consider using a hardware wallet. These offline devices securely store your private keys.
    - Software Wallets: If you prefer software wallets, choose reputable options and keep your computer and software up to date with the latest security patches.

5. **Private Key and Recovery Phrases Security**
   - Never Share: Never share your private keys or recovery phrases with anyone, under any circumstances.
   - Back Up Securely: Store backups of your recovery phases in a safe, offline location.

6. **Regular Monitoring is Critical**
   - Review Transactions: Regularly review your transaction history for any suspicious activity.
   - Update Software: Keep your operating system, browser, and cryptocurrency software up to date with the latest security patches.

7. **Report Scams**
   - Report Suspicious Activity: If you suspect a scam, report it to the relevant authorities (e.g., the FTC in the US).

Part IV
TAKE THE FIRST STEP ON YOUR CRYPTO JOURNEY

CHAPTER 11

FIVE WAYS TO EARN FREE CRYPTOCURRENCIES

CHAPTER OVERVIEW

This chapter shows five low-risk ways to earn small amounts of cryptocurrency while you learn. Rewards and availability change over time and by region, and there are real security risks—so start small, follow the safety checklist, and treat every 'free' offer with skepticism.

THE RISK-FREE WAY TO START YOUR CRYPTO JOURNEY

I have sat through many cryptocurrency presentations by highly intelligent and experienced presenters who delve into the history of Bitcoin and cryptocurrencies, explaining why they are good investments. These are excellent presentations that offer invaluable insights and good content. However, they don't provide beginners with the right tools or incentives to actually *start* using cryptocurrencies.

At the end of my presentations, I show the audience three ways they can earn cryptocurrencies. Invariably, several attendees pull out their phones during this part and take pictures of the free cryptocurrency slides. Some even download the apps during the presentations. I've found that this is the most effective way to get listeners and attendees to actually download wallets and start engaging practically with cryptocurrencies.

Earning free cryptocurrencies is the ideal, low-risk way to complete your first cryptocurrency transaction. Once you start earning cryptocurrencies and become more comfortable with the price swings, you can decide if you want to invest in cryptocurrencies. Free cryptocurrencies are also valuable to experienced cryptocurrency investors because they understand the value proposition of cryptocurrencies. Although these are free cryptocurrencies, they will appreciate in value just as purchased cryptocurrencies.

HOW I EARN FREE CRYPTO (AND HOW YOU CAN AS WELL)

You don't need to actively trade in or mine cryptocurrencies to get started on your crypto journey. If you watch videos on apps, use social media, shop online, stay at hotels, rent cars, eat fast food, or engage in other online transactions, you can earn free cryptocurrencies.

One weekend, I decided to track my everyday activities to see where I earned free Bitcoin and other cryptocurrencies. These are listed below. As you'll see, they are activities most people do every week, if not every day.

- I used the Brave browser (Brave) for all the actions below, and I was awarded Basic Attention Token (BAT) for using the Brave browser.
- I used Lolli for my hotel and rental car purchases and was awarded in Bitcoin.
- I used Publish0x to read social media posts and articles. I tipped the author in USDC for writing the article and in USDC for reading it.
- I also used Presearch for web searches, and I was awarded PRE-tokens.
- I watched lessons and completed quizzes on CoinGecko, earning cryptocurrency for successfully completing them.

So, in one weekend, I earned BTC, USDC, AMPL, BAT, and PRE — without buying any of them! Keep in mind that you will not earn a significant amount of cryptocurrencies through this process; however, as you continue to use these and other sites, you will earn cryptocurrencies without purchasing them. When the prices increase, the value of your free cryptocurrencies will increase as well, growing your overall crypto balance sheet. That's the benefit of earning free cryptocurrencies!

FREE CRYPTO DISCLAIMERS

Before you read the rest of this chapter and start your quest for free crypto, here are some important disclaimers.

- There is no guaranteed way to earn a significant amount of free cryptocurrency. Many methods require substantial time and effort for a relatively small potential reward.
- Some methods of acquiring free crypto may have hidden costs, such as transaction fees or the need for expensive equipment.
- Free crypto giveaways may require you to provide personal information or connect to untrusted websites, both of which pose security risks.
- Free crypto offers may come with complex terms and conditions. Read the fine print carefully before participating in any free crypto offers.

For all of the reasons stated above, research the company offering the free cryptocurrencies and understand the underlying business case before providing your personal information.

Now that we covered the disclaimers, here are the five ways I earn free cryptocurrencies:

1. LOLLI

Lolli is a rewards program that pays users in USD or Bitcoin for using the platform to purchase travel or items. Once you sign up, you can choose to receive your rewards in USD or Bitcoin; however, once you make a selection, you cannot change it. Of course, I chose Bitcoin, especially during the 2022/23 bear market, when it was trading at a 76% discount from its all-time high. This allowed me to receive free Bitcoin at a discounted price, so it will become more valuable when its price increases.

I use Lolli to purchase hotel rooms, rental cars, and other items from merchants. Lolli also has a 90-day locking period — you must wait 90 days before you can transfer the Bitcoin you have earned. Finally, you must earn $25 in Bitcoin before transferring it to a wallet.

How to use Lolli (3 steps):

1. Install the extension and enable pop-ups.
2. Click 'Activate' before checkout.
3. Track pending rewards → confirm after merchant returns window (can take weeks).

Note: availability varies by country; some rewards require account verification.

2. PUBLISHOX

Publish0x is an informational social media platform that rewards content creators and readers with cryptocurrency for using its services. When content creators write and post articles on Publish0x, readers can tip creators and themselves in cryptocurrencies. The site is a valuable resource for staying informed about the latest developments in blockchain and cryptocurrency, as well as public opinions on these updates. If you find a content creator you like, you can follow them on Publish0x.

Publish0x limits tipping yourself or others to once every 10 minutes. Currently, you can tip the creator and yourself in USDC. I say "currently" because historically, Publish0x has rewarded content creators and readers with other cryptocurrencies. Publish0x tips rotate among supported tokens; check the current token list and withdrawal minimums for details.

PubishOx will likely add other cryptocurrencies as tip options in the future as well.

I use PublishOx as a social media platform to earn cryptocurrencies by reading articles and posts on the platform. If you are interested in conducting social media research on cryptocurrencies or blockchains, PublishOx lets you save articles for future reference.

I have also written and published numerous articles on cryptocurrency and blockchain on PublishOx, and readers have tipped me in crypto. In fact, you can earn more crypto as a content creator than as a reader.

3. BRAVE BROWSER

The Brave browser (Brave) is a blockchain-based internet browser that pays users in BAT for surfing the web with it. Brave Rewards lets some users earn BAT for opting into privacy-preserving ads. Availability, payout rates, and supported assets vary by country and over time; please check the current terms before you start. I use it for most of my internet browsing. Currently, Brave rewards users with the following tokens: Solana (SOL), Ethereum (ETH), Basic Attention Token (BAT), Matic (MATIC), Binance (BNB), Avalanche (AVAX), Fantom (FTM), and Filecoin (FIL), but that may change in the future.

Note: KYC may be required to withdraw BAT via custodial partners.

4. PRESEARCH

Presearch is a decentralized search engine that rewards users with Presearch tokens (PRE) for browsing the web using the Presearch browser. You can also earn PRE by operating a node or by inviting others to join Presearch. As a node operator, you contribute to the operation of the search engine process. You register as a node operator, download the node software, and use the free space on your computer to power searches.

Presearch also offers keyword staking. Businesses can increase their website traffic by using Presearch keyword-stacking ads. Under this model, Presearch advertisers do not have to pay per click, impression, or action.

6. COINGECKO "EARN AND LEARN"

CoinGecko is a cryptocurrency data aggregator, which is a fancy way of saying that it collects and analyzes data from the internet and makes it available to users. CoinGecko occasionally runs 'Learn & Earn' campaigns (watch videos and pass quizzes). Campaigns are limited-time and region-dependent—check the current page for availability. The videos, quizzes, and tasks are updated regularly, so there is always an opportunity to earn more cryptocurrencies.

FIVE WAYS TO EARN FREE CRYPTO: EFFORT VS. REWARD VS. RISK

Method	What You Do	Typical Payout*	Wait Time	Region / KYC Notes	Top Risks	Official Page
Brave Rewards	Enable privacy-preserving ads in the Brave browser	BAT (~$1–3/mo casual browsing)	Monthly payouts	Some regions only; the custodial partner may require KYC	Low earnings; changing terms	brave.com/rewards
Lolli	Shop online at partner merchants	1–5% back in BTC	Weeks (after merchant return window)	US-centric; ID check for withdrawals	Missed activation; country limits	lolli.com
Publish0x	Read/write posts; split tips with author	Rotating tokens (under a penny/tip)	Immediate tip → min withdrawal	Token availability changes, payout threshold	Low liquidity coins	publish0x.com
Presearch	Search engine → earn PRE; advanced users can run nodes	Fractions of PRE per search	Daily accrual	Node operation requires staking + uptime	Set up errors; staking loss	presearch.io

Method	What You Do	Typical Payout*	Wait Time	Region / KYC Notes	Top Risks	Official Page
CoinGecko Earn & Learn	Watch short videos, complete a quiz	Small token drops ($1–3)	Campaign-based, limited-time	Region restrictions; often KYC	Fake links, phishing lookalikes	coingecko.com/earn

**Estimates as of 2025; subject to change. Always check current terms.*

SAMPLE WEEKEND EARNINGS LOG (TWO DAYS, 2025)

Activity	Token Earned	Approx. Amount	Time to Credit	Notes & Limitations
Brave browser ads (3 hrs. casual browsing)	BAT	0.35 BAT ($0.10)	End of the month	Very low but passive; region-dependent
Lolli purchase at Nike ($100 shoes)	BTC	0.0012 BTC ($1.00)	45 days (after return window)	Requires remembering to activate the extension; the merchant network is limited outside the US.
Publish0x (read + tip 5 articles)	VARIOUS (USDC)	~$0.05–0.10	Instant	Tokens are micro amounts; liquidity varies; payout threshold can delay withdrawals.
Presearch (20 searches while testing)	PRE	2 PRE ($0.06)	Daily accrual	Must diversify usage (not all searches credited); node staking optional.
CoinGecko "Earn & Learn" quiz (1 campaign)	MIXED (e.g., AAVE, SAND)	~$2–5	Within 24 hrs.	Campaigns are limited; they often require a quick signup before slots fill.

Total Weekend Haul (Illustrative): Approximately $2–$5 worth of cryptocurrency (at market rates).

SECURITY CHECKLIST SIDEBAR

Before you try any "free crypto" method:

Wallet Safety

- Use a separate wallet for experiments; keep larger holdings offline.
- Never store seed phrases in cloud drives or screenshots.

Transaction Hygiene

- Never blind-sign contracts or approve unlimited spending (do not set Approval For All).
- Periodically revoke token approvals using a trusted tool (e.g., revoke.cash).

Device & Browser

- Keep OS, browser, and extensions updated.
- Avoid mixing work accounts and experimental wallets in the same browser.

Phishing Red Flags

- Lookalike URLs (e.g., coingeck0.com).
- Urgency language ("Act now, claim instantly!").
- Spelling mistakes and mismatched logos.

Social Channels

- "X" and Discord are rife with impersonators. Verify official handles, never DM seed phrases.

Tax & Legal Note

- Rewards may be taxable income, and some platforms require KYC verification depending on your country.

RISKS OF RECEIVING FREE CRYPTOCURRENCIES

If it sounds too good to be true, it is. Most 'free crypto' scams exploit urgency and unfamiliar wallet permissions. Never blind-sign transactions, never share seed phrases, and regularly revoke token approvals.

Another reason there are only five recommendations is that I only earn free cryptocurrencies for my usual activities. Other ways to earn free cryptocurrency include watching ads or completing surveys. These methods can be time-consuming, offer minimal rewards, and expose you to junk emails. If you start watching ads, the algorithm may display more of the same or similar ads in your browser or app. If you start completing surveys and give your email address to complete the survey, you may begin to receive more spam.

PLAY TO EARN

Finally, playing games is another way to earn free cryptocurrencies. If you are a gamer, playing to earn (P2E) is a great way to earn cryptocurrencies or in-game assets — you'll earn free crypto while doing what you love! Players often purchase in-game items like characters, weapons, or land as NFTs. These assets can be used to progress in the game and generate cryptocurrency or tokens, which can then be traded or converted into fiat currency.

One major drawback of playing P2E games is that they often yield only a small amount of cryptocurrency. As such, players may spend a lot of time playing the game to earn as much as they can — in my opinion, the small amount of crypto earned is not worth the time.

Free cryptocurrencies can be tempting, but, as discussed in the disclaimers earlier in this chapter, there are risks to be aware of before attempting the five strategies mentioned. The best way to protect yourself and your cryptocurrency in the P2E world is to ensure you play on trustworthy P2E platforms; otherwise, you risk exposing your cryptocurrency or personal information.

FREE CRYPTOCURRENCY SCAMS

Many schemes promise free cryptocurrencies but are actually trying to steal your money or personal information. Be careful of anything that sounds too good to be true, especially if it involves:

- Sending any money upfront (even for "fees").
- Giving away your private cryptocurrency keys.
- Clicking on suspicious links or downloading unknown software.

THE DANGER OF WORTHLESS COINS AND AIRDROPS

Some of the biggest 'free crypto' payouts come from airdrops and testnet participation. Early Uniswap users received 400 UNI tokens in 2020 (worth over $1,000 at the time). Testnets like zkSync and Arbitrum reward users for trying beta features. These are speculative — not guaranteed — but highlight how early adoption can turn into windfalls. Of course, many free cryptocurrencies, especially those making airdrops, are worth very little when they are initially minted and dropped. Airdrops are a method where blockchain companies "drop" a few free cryptocurrencies or tokens into wallets when wallet owners complete specific marketing tasks, such as liking, following, and posting on social media, or participating in surveys or games.

Cryptocurrency companies often give away their cryptocurrencies for various reasons. First, they aim to develop brand recognition so that users will download the cryptocurrency wallet, use the coins, and hopefully start buying. Second, depositing free coins into users' wallets is reported to cryptocurrency exchanges as movement, and such use can increase the cryptocurrency's market capitalization and attract new investors.

In theory, this is a sound approach, except that some of these giveaways are scams or worthless coins that will never really appreciate in value. So, how can you tell the difference between a useless coin that has no value and a legitimate coin that may actually have value?

You can identify worthless cryptocurrencies using the same research and analysis process we have discussed in the previous chapters.

- Evaluate the project's whitepaper for technical feasibility and market viability.
- Assess the executive team's experience and track record.
- Scrutinize the tokenomics, including supply, distribution, and utility.
- Consider the overall market sentiment and trading volume.
- Beware of projects with unrealistic promises, excessive hype, or anonymous teams.
- Utilize online tools and resources for market data and analysis.
- Safeguard your personal information and be cautious of untrustworthy websites and other potential security threats.

This approach will help you distinguish between legitimate airdrops and worthless ones.

Beware: Free crypto is never risk-free. Watch out for scams disguised as airdrops; always double-check URLs. Remember, even $5 in rewards may be taxable.

SUMMARY

The best way to summarize this chapter is to provide strategies to help you do the research and safely earn free cryptocurrencies:

1. Thoroughly research the cryptocurrency company before claiming any free cryptocurrencies. Research the company website, check social media for reviews, and verify if the company is legitimate.
2. Scrutinize the reviewers' profiles for consistency. Fake accounts often have generic names, few followers, and limited activity. For example, I once saw reviews from accounts named Patricia, Patty, and Pat, with no last name. I knew right away it was a scam.
3. Analyze the Content: Look for detailed, informative reviews that offer both positive and negative aspects of the cryptocurrency. Overly enthusiastic or negative reviews without substance might be suspicious.
4. Verify and cross-reference the information with other reputable sources and social media sites. If the review contradicts widely accepted facts or data, it's probably a false review.
5. Beware of Paid Reviews! Some influencers or reviewers might be paid to promote certain cryptocurrencies. Be careful of overtly promotional content.
6. Look for red flags, including grammatical mistakes, misspelled URLs, excessive use of exclamation points, unrealistic promises, or pressure tactics. These are common signs of a scam.
7. Never share your private keys. Some scammers are bold enough to ask for your private key to complete a transaction or verify your account. Private keys are like the password to your cryptocurrency holdings. Never share them with anyone!
8. If you accept a free crypto offer, be very careful with any personal information you provide.

If you follow the recommendations in this chapter, you will start safely and securely earning free cryptocurrencies in no time! These rewards won't replace income — most are worth only a few dollars. But they're powerful 'hands-on tutorials.' By claiming even $5 in BAT or USDC, you learn how wallets, transfers, and compliance really work — lessons more valuable than the tokens themselves.

SUCCESS STRATEGY ACTION ITEMS

This chapter covers five recommendations to earn free cryptocurrencies, essentially serving as a chapter on success strategy action items. It is relatively easy to earn free cryptocurrencies using any one of the strategies in this chapter, but using all of them can be time-consuming. One approach is to select actions you perform regularly. This way, you earn free cryptocurrencies for performing your regular actions. However, I do have a few additional recommendations.

1. Many other sites offer free cryptocurrencies, but beware of platforms that scam you for your personal information. Be sure to research the free cryptocurrency site thoroughly on social media platforms, like Reddit and Discord, before divulging any personal information.
2. Social listening tools are invaluable in identifying potential scams in the cryptocurrency and NFT space. These tools monitor online conversations across various social media platforms, forums, and other digital channels. By analyzing the content of these conversations, you can uncover patterns, red flags, and early indicators of fraudulent activity. Examples of social listening tools include:

    - Hootsuite: A popular social media management tool that also includes social listening capabilities.
    - Mention: A dedicated social listening tool that provides real-time monitoring of mentions across various platforms.
    - Brandwatch Analytics: A powerful tool for analyzing large volumes of social data and identifying trends.
    - Talkwalker: A social listening platform that offers insights into consumer conversations and sentiment.

FREE CRYPTOCURRENCY QR CODES

1. **Lolli**

2. **Publish0x**

3. **Brave Browser**

4. **Presearch**

5. **CoinGecko Earn**

CHAPTER 12
HOW TO START YOUR OWN BLOCKCHAIN BUSINESS

"Smart companies will use blockchains to turn data into dollars, and trust into profits, for the business of tomorrow."

Eric Guthrie, Esquire

Author, *Blockchain or Die (2019) and Blockchain or Die Harder (2026)*

The final quote in this book is meant to inspire a forward-thinking business or professional approach to blockchain technology. Blockchain technology is not just a tool. It is the foundation for building the products and services of the future. Visionary CEOs don't just react to the present — they anticipate the needs of tomorrow and develop futuristic solutions that defy the limits of possibility. By leveraging the transformative power of blockchain, these leaders attract both cutting-edge investors and customers eager to embrace innovation. In this chapter, you'll discover how to join their ranks and turn your vision into reality and profits.

CHAPTER OVERVIEW

This chapter will focus on potential ways to create a profitable blockchain business, including some of the practices I used to develop blockchain profit centers in my business. It will also describe various professional services that can be used to build a career, create a blockchain business, or establish a blockchain profit center. Blockchain technology won't automatically turn data into dollars. But founders who combine vision with execution, create entirely new business models — from decentralized finance to tokenized real estate. The difference between hype and profit lies in identifying the right use case, building credibility, and navigating regulation. This chapter shows how to do exactly that.

ADVANTAGES OF BLOCKCHAIN TECHNOLOGY FOR SMALL BUSINESSES

There are many advantages for small businesses to use blockchain technology. Blockchain technology enables small and disadvantaged businesses (SDBs) to access capital more easily and efficiently by issuing their own cryptocurrencies or digital tokens through ICOs or STO processes. In short, SDBs can raise capital by selling portions of their assets.

In addition, when SDBs establish themselves as authorities on blockchain technology, they gain visibility in this technologically innovative field. This visibility attracts potential customers and investors and builds valuable partnerships. SDBs can position themselves as thought leaders, demonstrating a forward-thinking approach and ability to adapt to technological advancements.

Becoming a thought leader in blockchain technology can be a powerful strategy for SDBs to gain a competitive edge, attract valuable customers, and position themselves for future growth and success. Offering blockchain-related products or services can be a profitable niche market for SDBs.

ICOs were once a fast path to fundraising — until the SEC cracked down in 2018. Today, small businesses are increasingly turning to tokenized crowdfunding (Reg CF/Reg A+), DAO-driven micro-funding, or loyalty tokens. For example, CityDAO raised $2M by selling governance tokens tied to parcels of Wyoming land.[266] Small businesses can replicate this model in niche markets — but only with legal guardrails.

There are many success stories of small businesses utilizing blockchain technology, and the stories behind two very successful blockchain companies, Ethereum and Storj, are worth following.

THE ETHEREUM BUSINESS STORY

I first learned about Ethereum in 2016, on the same day I began researching Bitcoin. While Bitcoin sparked my initial interest in cryptocurrencies, Ethereum was the company and cryptocurrency that truly sparked my fascination and passion for blockchain technology. There were several reasons I was impressed with the company, but the most significant one was Ethereum's business model.

Ethereum made money programmable and introduced smart contracts. These concepts allowed Ethereum to raise funds, spin up new marketplaces, and test models that banks could never offer.

Ethereum began as a concept proposed by Vitalik Buterin in a white paper published in 2013, entitled "A Next Generation Smart Contract & Decentralized Application Platform," which changed business processes and the world of contractual arrangements.[267] Published five years after Satoshi's Bitcoin white paper, the Ethereum white paper built on Bitcoin's blockchain technology and arguably had a broader scope and greater impact on businesses, organizations, and governments. Ethereum outlined

---

[266] Hubbard, Sarah; Trivendi, Anand; and Sharma, Mayank. Case Profiles of Decentralized Autonomous Organizations. Harvard Kennedy School Belfer Center: Technology and Public Purpose Project Case Study. July 2023.

[267] Buterin, Vitalik. A Next Generation Smart Contract & Decentralized Application Platform. Ethereum.org, July 20, 2015. https://ethereum.org/whitepaper/#ethereum-whitepaper (Accessed December 1, 2025)

a vision for a decentralized platform that could execute smart contracts that were programmed with self-executing code.[268] Here's a chronological analysis of Ethereum's evolution:

1. In 2014, Ethereum conducted a highly successful crowdfunding campaign, raising over $18 million in Bitcoin. This funding provided the initial capital to develop the Ethereum platform.
2. The Ethereum network was launched on July 30, 2015, with the creation of the first block, known as the genesis block. This marked the official birth of the Ethereum blockchain.
3. In January 2017, "Etheroll" was launched on the Ethereum network. Ethereum's first-ever dApp was a groundbreaking project in many ways, marking the beginning of a new era in blockchain technology. Etheroll was designed to provide users with a unique and decentralized way to gamble transparently and fairly using Ether (Ethereum's cryptocurrency).
4. DappRadar.com reports that as of June 2025, Ethereum has over 5000 dApps.[269]
5. As of June 18, 2025, Ethereum is worth approximately $270 billion.[270]

Ethereum showed the world that money itself could be programmable. That one breakthrough unlocked a flood of experiments — from fundraising to digital services.

Ethereum is a huge success story for small businesses, but it is not the only one. The story of Storj is another brilliant example of the success of a visionary small business.

---

[268] Buterin, Vitalik. Ethereum Whitepaper. Ethereum.org, Updated October 27, 2025. https://ethereum.org/whitepaper/ (Accessed November 2, 2025)

[269] Top Blockchains, https://dappradar.com/rankings/chains (Accessed October 28, 2025)

[270] Ethereum. https://crypto.com/price/ethereum (Accessed October 28, 2025)

THE STORJ BUSINESS STORY

A few months into researching cryptocurrency companies, I discovered Storj and was impressed by their innovative business model. Storj is a decentralized cloud storage business. Instead of renting a giant data center like Amazon, Storj lets anyone rent out unused hard drive space and get paid in tokens. For entrepreneurs, the lesson is significant: blockchain enables markets for underutilized assets. I was even more impressed by the company's origin story. Shawn Wilkinson conceptualized Storj after attending a Texas Bitcoin Hackathon in 2014 as a Morehouse College student.[271] Here is a chronological analysis of Storj's evolution:

1. In 2014, Storj's first token sale raised half a million dollars.
2. Between 2015 and 2017, they raised another $3 million, which enabled them to launch the second iteration of their product.
3. Storj launched its second token sale in 2017 and raised $30M in 6.5 days.
4. In 2018, Storj formed a strategic partnership with Google Cloud, integrating its technology into the Google Cloud Platform. This collaboration increased Storj's visibility and credibility within the enterprise market.
5. As of June 2025, Storj has a market cap of $92 million.[272]

Driven by its innovation to revolutionize the cloud storage industry, Storj epitomizes the groundbreaking potential of small blockchain businesses. Storj succeeded not only because it decentralized storage, but

---

[271] (No Author) Shawn Wilkinson – A Founder Story. AtlantaTechVillage.com, April 18, 2018. Shawn Wilkinson conceptualized Storj after attending a Texas Bitcoin Hackathon in 2014 as a Morehouse College student (Accessed November 2, 2025)

[272] Storj. https://crypto.com/price/storj (Accessed October 28, 2025)

also because it struck early enterprise partnerships (Google Cloud). Competitors like Sia had similar tech but lacked Fortune 500 partnerships.

These blockchain success stories are already in the history books. The next session will discuss how you can build your blockchain success stories.

CREATING BLOCKCHAIN PROFIT CENTERS

When I first stepped into the blockchain industry, I wasn't building a cryptocurrency or a blockchain enterprise solution; I was training on best practices and consulting on related legal issues. The point is that you don't have to code to profit. If you can explain, translate, guide, and provide value in this ecosystem, you can create a blockchain profit center. I define "blockchain profit centers" as the part of a company that has a blockchain-related product or service with specific goals, revenue, and costs. Creating a profitable blockchain-based business is more challenging than establishing a traditional one.

Currently, most people do not understand blockchain technology, and even fewer know how to use it. This lack of understanding presents a major business development challenge. As a result, blockchain businesses may need to take additional steps to educate potential customers before they can effectively sell their products or services, incurring extra time and expense to source clients and close deals.

SOURCING CLIENTS

How do you source potential clients for your blockchain profit center? Many marketing strategies can attract potential clients to your blockchain profit center, including blogs, vlogs, podcasts, training, NFTs, and, in my opinion, publishing high-quality books. All of these examples are excellent educational tools, but more importantly, they are effective marketing and branding tools. I chose to write '*Blockchain or Die*' as I was already a bestselling author, and publishing a well-researched and well-written blockchain book was a great way to help educate a wider demographic about blockchain technology and establish myself as a blockchain expert. Additionally, online book sales generate passive income. Published books like *Blockchain or Die* are profit centers that generate more profit centers. The following sections will show how I made blockchain profit centers for my company, Better ME Better WE.

PAID SPEAKING ENGAGEMENTS

A book is not just a vanity project; it's a lead funnel. My book generated 5X ROI in consulting leads compared to direct sales. Speaking gigs added another 20% of revenue streams. The lesson for founders: treat content (books, webinars, podcasts) as a profit center multiplier — not just a brand-building tool. When my first book, *Diversify or Die,* was published in 2016, I started receiving paid speaking engagements. The number of these offers and engagements increased considerably in 2019, following the publication of *Blockchain or Die*. Speaking engagements are a significant source of new blockchain clients as they provide brand exposure and credibility as an expert, allowing me to sell my blockchain products and services.

In addition to speaking engagements, offering workshops, webinars, or virtual events can create an additional revenue stream. You can also capitalize on your speaking engagements by selling related products or services, such as books, courses, consulting services, or merchandise.

TRAINING PROGRAMS

Cryptocurrency and blockchain training programs form a significant profit center for Better ME Better WE. Before entering the cryptocurrency space, I was a Certified Training Executive with over 20 years of experience in managing and delivering global training. This background proved highly transferable to the blockchain and cryptocurrency space.

When I first encountered cryptocurrencies and blockchain technology, it became clear that there was a pressing need for quality training in the field. In 2019, a popular training website offered approximately 100 courses on blockchain and 100 courses on cryptocurrency. However, as cryptocurrencies and blockchains became more competitive, the number of online training programs increased dramatically. As of December 2023, the same website listed 1,724 cryptocurrency and 2,537 blockchain courses. As of January 2025, there are 2330 cryptocurrency and 2933 blockchain courses on the same website.

However, creating training courses was only part of the equation. Monetizing them required effective marketing to drive registrations. This aspect was new to me as I had limited experience in sales and marketing. I applied a foundational marketing rule I had learned many years ago: give away something for free to build your brand and business.

I started offering free webinars to the public and various organizations. These webinars served two purposes: promoting *Blockchain or Die* and selling blockchain and cryptocurrency training. Free webinars can be a powerful tool for businesses, offering a range of benefits:

- Free webinars are good lead generators. They attract potential customers who are interested in a particular industry or niche by offering valuable content. Capturing registration information enables follow-up and fosters relationships with these potential leads.
- Free webinars create brand awareness by showcasing expertise and establishing the company as a thought leader in its field. This can

significantly boost brand recognition, especially when reaching a wider audience than traditional marketing methods.

- Compared to in-person events, free webinars are an affordable way to reach a large audience. There are minimal overhead costs, allowing businesses to target a specific demographic.
- Through free webinars, businesses can connect with potential customers worldwide, making them a valuable tool for finding global clients and expanding into new markets.

The effectiveness of webinars significantly contributed to the success of *Blockchain or Die*, which was ranked as an Amazon #1 Bestseller on seven occasions. Following webinars with large audiences, it was common for attendees to purchase the book on Amazon. Consequently, the book would frequently attain the Amazon #1 bestseller status the following day.

The methods I employ to craft engaging and effective webinars are as follows:

1. I customize each webinar to address the audience's needs, interests, and knowledge level.
2. I provide a clear outline of my webinar's topics and the time allotted for each. This helps keep the audience engaged and focused.
3. I enhance my presentation with slides, videos, or other visual aids, which makes the content more engaging and easier to understand.
4. I facilitate interaction and audience engagement by asking questions, conducting polls, or using chat features.
5. Everyone loves free stuff — I give away knowledge on how to earn free cryptocurrencies.

These webinars not only boosted book sales but also promoted my other services. The next section will cover my consulting services.

CONSULTING

Like most consultants, it took time for my consulting business to become profitable. I began with DEIA consulting, but the DEIA market was saturated with competitors. Every successful consultant has an approach that sets them apart — to distinguish myself from other consultants, I wrote and published *Diversify or Die.* It helped my consulting business, but did not deliver the success I sought. Enter blockchain technology and cryptocurrencies.

Not all blockchain service niches are equal. Security audits are the most lucrative (with an average smart contract audit costing $50K–$100K), but they require great technical skills. There are many marketing agencies, so the failure rate is high due to competition and a lack of credibility. Supply chain consulting is underserved, especially in emerging markets. The savvy entrepreneur selects a niche with high demand, low supply, and favorable regulatory conditions.

Leveraging my 20 years of consulting experience, I entered the relatively untapped blockchain space. As an early adopter, there were very few blockchain consultants, so I was able to attract new clients. By using blockchain training programs as a feeder pool for my consulting services, I created a successful business model.

Consultants primarily serve in an advisory capacity, offering expert advice, recommendations, and solutions to clients based on their expertise and experience. I am often hired as a consultant to analyze blockchain integration challenges and provide strategic insights to address them.

Many blockchain companies, especially start-ups, need competent consulting that can:

- Increase access to funding and profitability.

- Improve marketing.
- Drive efficient operations and enhanced technological options.

Given the evolving nature of blockchain technology, consulting also involves conducting research to provide insights and solutions for unique blockchain challenges. These services help organizations assess the feasibility of blockchain adoption, design a blockchain strategy, identify and develop the most effective use case, and develop a profitable blockchain solution.

PROJECT MANAGEMENT

I also leverage my project management experience as a blockchain profit center. While consulting focuses on strategic advice, project management is about execution. Consultants are typically subject-matter experts in their fields, using their knowledge to make informed recommendations and proposals. However, they do not normally have a direct role in implementing or executing the solutions they suggest.

Project managers work closely with clients but primarily coordinate project planning and execution. By overseeing the project from initiation to completion, project managers also focus on the practical aspects of execution, such as defining tasks, allocating resources, managing timelines, and ensuring objectives are met within budget and schedule.

LEGAL OR REGULATORY COMPLIANCE

Finally, all the other services function as a feeder pool for my blockchain legal services. Many clients either heard me speak at a conference or attended one of my blockchain training sessions before seeking my legal advice. On numerous occasions, I have first served as a consultant and later transitioned to provide legal advice on blockchain technology or cryptocurrency.

As discussed in this book, blockchain technology raises numerous legal and regulatory issues. Legal professionals with expertise in blockchain technology can help businesses navigate the legal landscape, ensuring compliance with applicable laws and regulations, including:

- Assisting in the formation of legal entities, choosing the appropriate corporate structure, and ensuring compliance with corporate governance requirements.
- Drafting and negotiating partnership, investor, and stakeholder agreements.
- Engaging with regulatory authorities on behalf of blockchain companies to provide input on regulatory frameworks.
- Handling any legal disputes that may arise, including contract disputes and disputes with cryptocurrency exchanges.
- Assisting with obtaining necessary licenses and permits, such as money transmittal licenses, for operating blockchain-based platforms in various jurisdictions.
- Establishing compliance protocols to mitigate legal risks.

While legal advice is crucial to the operations and efficiency of blockchain businesses, companies' legal needs vary. For example, legal services for blockchain technology companies play an important role in navigating the complex and rapidly evolving regulatory landscape. However, every blockchain company must consider and resolve various types of legal issues.

Here are three types of blockchain technology companies and the type of legal issues they should consider and resolve:

1. **Companies that issue security tokens or NFTs**

Legal services are necessary to ensure that token classification (e.g., utility tokens, security tokens) and sales comply with the applicable regulations.

Legal professionals can also advise on compliance measures to mitigate money laundering risks and address intellectual property issues related to NFTs.

2. **Identity management companies with unique blockchain software and protocols.**

These companies require legal advice on compliance with data protection regulations, such as the General Data Protection Regulation (GDPR), to protect user privacy. Legal professionals can also help with filing copyrights, trademarks, and patents, enforcing contracts to protect intellectual property rights, and developing training programs for employees to ensure awareness of legal and regulatory obligations.

3. **Decentralized autonomous organization/company**

Legal professionals can review and audit smart contracts to ensure compliance with legal requirements and industry standards, minimizing vulnerabilities and potential disputes.

Engaging with legal services tailored to the specific needs of blockchain technology companies is essential for ensuring legal compliance, minimizing risks, and fostering a conducive regulatory environment that promotes innovation in the blockchain space.

OTHER BLOCKCHAIN PROFESSIONAL SERVICES

Blockchain technology has gained significant attention and adoption across various industries, offering a diverse range of professional career options. You can create a business that supports industries with their blockchain solutions or products. The following business or career opportunities are possible options for individuals with the right skill set and an interest in blockchain technology.

BLOCKCHAIN DEVELOPMENT SERVICES

Blockchain developers create and maintain blockchain networks, smart contracts, and decentralized applications (DApps). They can also help build custom blockchain solutions. For example, smart contract developers design, develop, and audit smart contracts to ensure they function correctly and securely.

BLOCKCHAIN INTEGRATION SERVICES

Integrating blockchain technology into existing business infrastructure can be complex. Blockchain integration services encompass solutions and tools that enable the seamless incorporation of blockchain technology into existing business systems and processes. These services help facilitate the seamless incorporation of blockchain technology into existing business systems and processes. Blockchain services enable organizations to leverage blockchain technology in various areas, including security, transparency, and trust, by integrating it with their existing products, services, software, data, policies, and procedures.

BLOCKCHAIN SECURITY AUDITS SERVICES

Blockchain security experts assess the security of blockchain networks and smart contracts, identify vulnerabilities, and provide recommendations to mitigate risks.

For example, blockchain audits are crucial for identifying vulnerabilities, detecting smart contract bugs, and preventing financial loss and reputational damage. Security audits also uncover potential security loopholes, such as unauthorized access or data breaches, safeguarding user information. By assessing consensus mechanisms, audits can mitigate the risks of network attacks. Finally, audits verify the accuracy of transaction data, helping prevent fraud and ensuring transparency.

Ultimately, regular audits bolster trust and confidence in the blockchain system.

If your blockchain software fails, it will severely damage your company's brand, threaten profits, and risk company failure.

TOKENIZATION SERVICES

Tokenization services involve creating digital tokens that represent assets or rights on a blockchain or distributed ledger system. Tokens can represent various assets, including traditional currencies, commodities, real estate, company shares, or even unique digital assets, such as in-game items or art pieces. Tokenization experts help navigate the token creation process to the point where it is actually used as a utility, reward, or security token.

BLOCKCHAIN MARKETING AND PUBLIC RELATIONS SERVICES

Marketing and public relations are essential factors for blockchain solutions that require users to download wallets and tokens. After all, if potential users are unaware of the token, they cannot download the wallet and purchase it. If you're launching a blockchain-based project or cryptocurrency, marketing and PR experts can help create and execute efficient strategies to reach your target audience.

BLOCKCHAIN SUPPLY CHAIN SERVICES

Blockchain supply chain experts essentially act as project managers to coordinate the integration of a blockchain solution into a company's supply chain. This entails coordinating all supplier manufacturers, transport companies, storage companies, retailers, and any other company in the supply chain.

Blockchain or Die reported that Walmart used the IBM Blockchain Solution and created the IBM Food Trust Solution to digitize the food supply chain process, reducing the tracking time frame from seven days to an estimated 2.2 seconds.[273] Professionals specializing in blockchain for supply chain management can help companies implement transparent and traceable supply chain solutions using blockchain technology.

BLOCKCHAIN ANALYTICS AND DATA SERVICES

Blockchain analytics and data services involve collecting, analyzing, and interpreting data related to blockchain transactions and activities. These services leverage the transparency and immutability of blockchain technology to provide insights into cryptocurrency transactions, addresses, and smart contract interactions. They give the company the data it needs to measure the performance of its blockchain solution. Once the data is collected and analyzed, the company can make informed changes or improvements to its blockchain solution.

MY PREDICTIONS FOR THE FUTURE OF BLOCKCHAIN BUSINESSES

The next chapter for blockchain businesses is yet to be written, but there will definitely be a next chapter. So, how will the story unfold? That is the billion Bitcoin question.

Three forces loom over blockchain's future: artificial intelligence, quantum computing, and global regulation.

---

[273] Miller, Ron. Walmart is Betting on the Blockchain to Improve Food Safety. TechCrunch, September 24, 2018. https://techcrunch.com/2018/09/24/walmart-is-betting-on-the-blockchain-to-improve-food-safety/#:~:text=Save%20$450%20on%20your%20TechCrunch%20All%20Stage%20pass&text=After%20working%20with%20it%20for,standards%2C%20according%20to%20the%20company. (Accessed October 28, 2025)

AI and blockchain make an odd pair, but one with potential. AI can detect fraud or optimize network throughput; blockchain can secure training data on tamper-proof ledgers, creating auditable AI models. Experiments like SingularityNET are already building decentralized AI marketplaces, though adoption is still thin.

Quantum computing poses the darker challenge. Algorithms like Shor's could crack Bitcoin's elliptic-curve cryptography, exposing private keys. Grover's algorithm weakens hashing. The good news is that post-quantum cryptography is in development. NIST is already vetting lattice-based schemes, such as Kyber and Dilithium, and some blockchain teams are beginning early integration. However, the timeline is uncertain — a decade could pass before quantum computing becomes truly dangerous, or the breakthrough could come sooner.

Then there is regulation. It's not a single story but a patchwork: the EU's MiCA framework sets clear guardrails, while the U.S. remains fragmented between the SEC and CFTC, and China bans public crypto but invests heavily in state-backed blockchain (BSN). Singapore and the UAE, by contrast, run pro-innovation sandboxes. The future of blockchain won't be decided globally at once; it will be shaped by regulatory competition between regions.

In short, blockchain's trajectory is uneven. It will likely accelerate in finance, supply chains, and gaming, while lagging in government and healthcare. Its future is neither guaranteed nor doomed — it will depend on how well the ecosystem meets its toughest tests: scalability, interoperability, trust, and compliance in a quantum-ready world.

By anticipating upcoming trends, businesses can develop strategies to capitalize on opportunities, mitigate risks, and gain a competitive edge by being the first to introduce new products or services or adapting their business models to meet changing consumer demands. Here are my predictions for the next chapter in the blockchain business.

INCREASED COMPETITION AND CONSOLIDATION

As the blockchain industry matures, competition will intensify, and larger, better-funded companies will begin acquiring smaller companies with unique and successful technology applications.

INTEGRATION WITH ARTIFICIAL INTELLIGENCE (AI)

Blockchain technology and AI can be combined to create more intelligent and efficient applications, such as predictive analytics and fraud detection. Blockchain technology can also provide a secure and immutable foundation for AI-driven data management. Additionally, AI and blockchain technology are already live, with notable examples including Fetch.ai (autonomous agents), Ocean Protocol (tokenized data for AI), and SingularityNET (an AI marketplace). These aren't hypotheticals. They show that entrepreneurs who combine secure data with AI analytics can immediately target healthcare, supply chain, and energy.

BLOCKCHAIN TECHNOLOGY AND QUANTUM COMPUTING

Quantum computing leverages the principles of quantum mechanics to perform calculations exponentially faster than traditional computers. This is achieved through qubits, which can exist in multiple states simultaneously. Quantum risk is real as well. Google's Sycamore has already demonstrated quantum supremacy in narrow tasks.[274] NIST is drafting post-quantum cryptography standards (PQC) that are expected

---

[274] Orf, Darren. Scientists Say We've Finally Reached Quantum Supremacy. For Real This Time. Popular Mechanics, September 29, 2025. https://www.popularmechanics.com/science/a68068223/quantum-supremacy-breakthrough/ (Accessed November 5, 2025)

to be finalized by 2027.[275] The entrepreneurial lesson: if you build blockchain businesses now, ensure they're quantum-resistant by design — otherwise your security model has an expiry date. Since quantum computing is a relatively new technology, the following is a brief example that illustrates how it works.

THE QUANTUM COMPUTING LIBRARY STORY

Imagine a magical library that is accessible to anyone in the world. In a regular computer library, such as on your phone or laptop, you are looking for a specific book. Each book has a title, and each title is either present (a "1") or absent (a "0"). To find your book, you have to look through them one by one. If you want to find a very rare book, it may take a considerable amount of time, checking each title in sequence. It's like searching for a needle in a haystack, but instead of checking each piece of hay, your magic lets you instantly know which piece of hay is *most likely* the needle, and all the non-needles simply vanish.

The Quantum Computer Library features three groundbreaking processes to expedite and simplify this complex and lengthy process: Ghost Books, Telepathic Books, and the Echo Chamber. Let's look at all three.

1. "Ghost Books" (Superposition): Instead of books being just "there" or "not there," in this quantum library, books can be "ghost books." A ghost book is like a shimmering, blurry image of many different books all at once. It's all the books existing simultaneously. These are your qubits.

[275] News. NIST Selects HQC as Fifth Algorithm for Post-Quantum Encryption. NIST.gov, March 11, 2025. https://www.nist.gov/news-events/news/2025/03/nist-selects-hqc-fifth-algorithm-post-quantum-encryption#:~:text=NIST%20has%20chosen%20a%20new,resistant%20algorithm%20called%20ML%2DKEM (Accessed December 2, 2025).

- So, if you have three ghost books, they aren't just representing "Book A, Book B, Book C" or "Book A, Book A, Book C." They can simultaneously represent *all possible combinations* of books at once. This means the quantum computer can explore many solutions to a problem *simultaneously*, not just one after another.

2. "Telepathic Books" (Entanglement): Some of these ghost books are "telepathic." If you touch one telepathic ghost book, its shimmering state instantly affects and is known by another telepathic ghost book. They're linked in a way that regular library books aren't.

    - This "telepathy" or entanglement allows the ghost books (qubits) to communicate and share information in incredibly complex ways, letting them work together on a problem far more efficiently than if they were working alone. They can form very intricate relationships.

3. "Echo Chamber" (Interference): When you finally want to find the *real* answer (your specific book), the quantum library uses a magical "echo chamber." As all the ghost books explore different possibilities, the "right" answers begin to create strong, positive echoes, growing louder and clearer. The "wrong" answers, however, create negative echoes that cancel each other out, fading into silence.

    - This is called interference. The quantum computer carefully designs its "echo chamber" so that the correct solution's "echoes" reinforce each other, making it stand out brightly, while incorrect solutions' "echoes" cancel each other out, making them disappear.

Although this may sound like a Harry Potter Hogwarts library, this story actually describes the quantum computing process in non-technical terms.

THE QUANTUM ANSWER

Finally, when you ask the quantum library for the answer, all the "ghost books" collapse into their single, definite state, revealing the correct book (the solution) that was amplified by the echoes.

This is groundbreaking and powerful because it's not trying possibilities one by one. It's trying *millions or billions of possibilities simultaneously* using its ghost books and telepathy, and then using the echo chamber to highlight the correct answer. This enables it to solve certain types of problems so complex that even the fastest regular supercomputers would take billions of years to figure out.

Now, let's take the Quantum Library into the real world.

QUANTUM COMPUTERS AND CRYPTOGRAPHIC ALGORITHMS

Quantum computers could potentially break the existing cryptographic algorithms used by blockchains, compromising the security of many blockchains. To address this potential security issue, future blockchains may need to utilize quantum computing to develop more secure cryptographic algorithms, thereby making blockchain transactions even more resistant to hacking, particularly by quantum computers. These algorithms could even accelerate blockchain consensus mechanisms, leading to faster transaction times. Finally, quantum cryptography can provide advanced privacy features for blockchain transactions.

The cost of setting up quantum computing systems has to dramatically decrease before their potential can be realized in everyday business solutions. However, with the ever-narrowing gap between the conceptualization and implementation of new technologies, this marriage between blockchain and quantum computing may not be far off.

SUMMARY

This chapter exemplifies the book's core message: ***No Action Means No Profit***. It discusses how I used my experience and skills to create a blockchain profit center, and how you can do the same. Validate whether your blockchain use case solves a real inefficiency. Example: Supply chain pilots often fail because the bottleneck is transparency and not data entry. Fixing garbage in/garbage-out is the real problem.

If I had to summarize my cryptocurrency and blockchain journey into five steps, they are as follows:

1. **Build a Strong Foundation**: I started by understanding the basics of blockchain, cryptocurrencies, and smart contracts. I read white papers, especially those related to Bitcoin and Ethereum, as well as blogs, articles, videos, and podcasts.
2. **Develop and Incorporate Professional/Technical Skills:** I incorporated my legal and training skills into the crypto and blockchain space, and so can you. Don't just hire Solidity devs — hire compliance advisors. Reflect on how you can incorporate these into the blockchain space.
3. **Network Strategically:** I joined blockchain organizations, attended industry conferences, meetups, and webinars. Connect with blockchain professionals on social media, including: 𝕏, Discord, and Telegram.
4. **Gain Practical Experience**: I volunteered and contributed to cryptocurrency and blockchain projects. If you cannot find a project, start your own. Be sure to seek guidance and advice from more experienced blockchain professionals. To gain legal experience, I began representing my first clients at a discounted rate. After gaining sufficient legal experience, I began charging full price for my legal services.

5. **Continuous Learning**: Blockchain technology is rapidly evolving, so I stay updated with industry trends, new technologies, and regulatory changes. I also continually update my training and speeches to reflect the updates in this industry. *Blockchain or Die Harder* is one big update that I wrote to stay relevant.

I will end this chapter with two questions for your consideration:

1. For the entrepreneur, what skills, education, or training do you possess that can be leveraged to create a blockchain profit center?
2. For the career-minded individual, what skills, education, or training do you possess that can be leveraged to build a career in blockchain?

Once you answer these questions, you have taken the first step in your professional blockchain journey.

SUCCESS STRATEGY ACTION ITEMS

If you are considering starting a blockchain business, these may be the most essential Success Strategy Action Items in this book.

1. **Identify a Niche or Use Case**: Determine the specific industry for which your blockchain solution is tailored. Some industries include: finance, supply chain, healthcare, real estate, and other sectors where blockchain can deliver value.
2. **Market Research:** Conduct thorough market research to identify the demand for your blockchain solution. Understand your target audience, competition, and potential customers' pain points.
3. **Create a Business Plan**: Develop a comprehensive business plan outlining your vision, goals, revenue model, marketing strategy, and financial projections. A well-structured business plan will be crucial for attracting investors and partners.

4. **Legal and Regulatory Compliance:** Ensure that your blockchain business complies with the legal and regulatory requirements in your region. Consult legal experts to navigate any complex compliance issues.
5. **Build a Skilled Team**: Assemble a team with technical expertise in coding, cybersecurity, legal, and business management.
6. **Select the Right Blockchain Platform**: Choose the most suitable blockchain platform for your project, considering factors such as scalability, consensus mechanism, and developer community support.
7. **Security and Compliance:** Implement robust security measures to protect your blockchain system from hackers and cyber threats. Stay informed about new cyber threats to safeguard blockchain data.
8. **Development and Prototyping**: Develop a minimum viable product (MVP) or prototype to demonstrate the feasibility and functionality of your blockchain solution. This will be crucial for attracting investors and early adopters.
9. **Funding**: Secure funding for your blockchain business. Options include self-funding, venture capital, angel investors, crowdfunding, or blockchain-specific fundraising methods, such as initial coin offerings (ICOs) or security token offerings (STOs).
10. **Development and Testing**: Build your blockchain solution, be it a decentralized application (dApp), smart contract, or more complex network. Remember to test the technology to ensure its reliability and security.
11. **User Acquisition and Adoption**: Develop a user acquisition strategy to onboard customers and users onto your blockchain platform. Consider offering incentives and rewards to early adopters.
12. **Marketing and Promotion**: Create a marketing plan to promote your blockchain business. Utilize digital marketing, content

marketing, and community building to raise awareness and attract users.

13. **Launch and Monitor**: Launch your blockchain solution and monitor its performance. Be prepared to make adjustments and improvements based on user feedback and real-world usage. Track user adoption and interface to determine how you can improve the user experience.

Every wave of tech opens doors for entrepreneurs. With blockchain, the doors are wider than most people think — because the industries are still being written.

CHAPTER 13

BLOCKCHAIN OR DIE HARDER: CLOSING THOUGHTS

There are two parts to this revolutionary technology journey: the blockchain side and the crypto side. Although blockchain technology enabled cryptocurrency, it also created two interrelated but distinct industries. For example, most blockchain solutions do not use cryptocurrencies.

On the cryptocurrency side, we've journeyed through myths, markets, meltdowns, and momentum. From Bitcoin's white paper to El Salvador adopting Bitcoin as its fiat currency, from the SEC's courtroom showdowns, to Nigeria's eNaira. One thing is clear: cryptocurrencies are creating new and exciting ways to engage in transactions.

On the blockchain side, we have explored key protocols, technology updates, innovative approaches to corporate governance, real-world use cases, and emerging threats to blockchain security. Once thing is clear: blockchain technology is solving many previously unsolvable corporate and government problems.

We've come to the end of our journey, but this isn't a conclusion; it's an invitation to a new beginning. When we started, the title *'Blockchain or Die '* was a bold statement, a challenge to confront a world in flux. Now, as we stand at the end of these pages, you are no longer a passive observer but an informed participant ready to make your choice.

Blockchain or Die is more than a title; it is a declaration. It is the story of a technology that is fundamentally changing our world, and a challenge for you to choose your place within it. We have come to the end of this book, but your journey is just beginning. The knowledge you have gained is a key—a key to a new frontier of innovation, financial freedom, and creative expression.

The time for passive consumption is over. The lessons learned from each chapter, the insights into the market's evolution, the understanding of its applications, and the wisdom on how to navigate its risks—all of this now belongs to you. But knowledge alone is not enough. Success in this new paradigm is built on action. It's time to take out your compass and your roadmap, and use the **Success Strategy Action Items** that concluded each chapter as your guide.

Go back to those action items. Re-read them. Choose the ones that resonate most with your goals, whether that's investing, creating, or building. Use them to establish your first crypto wallet, to research your first NFT, to craft a business plan for a new blockchain venture. Let the spirit of these action items ignite your path. The future is not a destination we arrive at, but a ledger we write, block by block. And now, armed with this knowledge, it's your turn to mint the next chapter.

CLOSING LINE

Every era leaves a message for the next. Here's ours:

Don't just watch the future happen. Build it.

AFTERWORD

Congratulations on completing *Blockchain or Die Harder*. I hope this book provided you with many valuable insights and takeaways. One of the most important themes in both *"Blockchain or Die" books is to start learning, get involved, and, most importantly,* take action.

Many people tell me they are "planning" to invest in cryptocurrencies or "researching" the best ones. The problem is that they have been using the same lines and the same excuses for years. They are on the sidelines, too afraid to actually step onto the field. You will not make any money or build a cryptocurrency or blockchain business from the sidelines. You must take action.

Blockchain or Die Harder gives you all the tools you need to take action: technical, business, and legal information, ways to earn free cryptocurrencies, Success Strategy Action Items, QR codes, and websites. If you want the time you've invested in *Blockchain or Die Harder* to offer any returns, use the information and take action.

The afterword in *Blockchain or Die* asked: "What are you going to do with the information you learned?" This question persists, but with even higher stakes. Since 2019, there have been an increasing number of cryptocurrency exchanges, cryptocurrencies, and cryptocurrency and blockchain companies and projects, as well as more governments and corporations implementing cryptocurrency and blockchain solutions. They are taking action. Now it's your turn.

I look forward to seeing you prosper in the new blockchain economy.
Eric L. Guthrie, Esq.

ABOUT THE AUTHOR

ERIC L. GUTHRIE, ESQ., is an experienced global trainer who has conducted training sessions on blockchain, legal, supervisory, teamwork, and diversity in over ten countries.

"Blockchain or Die Harder" is Eric's third book. His other published books are:

- *Diversify or Die* (2016) Finalist for the 2017 American BookFest Best Business Book Award, 2018 Readers' Choice Award.
- *Blockchain or Die* (2019) Amazon #1 Bestseller seven times, 2020 NYC Big Book Award Winner, 2020 Independent Press Award Distinguished Favorite, 2020 Indie Book Awards Finalist, 2020 American BookFest Finalist.

Eric has spoken on blockchain technology and cryptocurrencies at over 30 in-person conferences and 100 virtual conferences.

Eric has been featured in numerous publications, including:

- ABC Money
- Markets Herald
- The Lincoln Citizen
- Washington Mail
- Hudson Weekly
- The Belmont Star
- Heart and Soul

Eric is the Director of Training for the Government Blockchain Association (2017–present) and a Partner at the Cogent Law Group (2018–present), a law firm specializing in FinTech, blockchain, and cryptocurrency practices. He is also an Adjunct Professor of Law, a Certified Training Executive, and a Certified Diversity Executive, bringing his training and writing skills to the cryptocurrency and blockchain world. He partners with numerous businesses and universities in several countries to host blockchain conferences. Learn more about Eric's services at BlockchainorDie.org or EricGuthrie.com.

INDEX

C

D

E

F

G

H

I

J

K

L

M

N

O

P

Q

R

S

T

U

V

W

X

Y

Z

Made in United States
Orlando, FL
04 November 2026

90208346R00236